PRAISE FOR
WHEN WOMEN WIN

"Ellen Malcolm shares the untold story of how women of color began to break into the top levels of American politics. An absolute must-read if you want to know how diversity can make Congress work for all of us." — **Ben Jealous, former president and CEO of the NAACP**

"This is the inspiring, fascinating story of the most potent women's political movement in the modern era — now 3 million members strong, built one woman at a time — that effectively broke up the old-boys club in American politics and paved the way for the kind of grassroots fundraising that made our government more democratic."
— **Howard Dean, former governor of Vermont and chair of the Democratic National Committee**

"A lively, fast-paced history of the influential political action committee that helps elect pro-choice, Democratic women. Drawing on interviews with Hillary Clinton, Nancy Pelosi, Elizabeth Warren, and others, she highlights the organization's impact on elections over the past thirty years . . . The book shows that EMILY's List's biggest contribution may be to make women in Congress seem so familiar that voters are now far more likely to judge women on their records and character than on their gender." — *Publishers Weekly*

"Prose as polished as a campaign speech . . . [A] narrative of sweeping change for women." — *Booklist*

"An inspiring portrait of a gutsy activist who produced a transformation in the political landscape." — *Kirkus Reviews*

WHEN WOMEN WIN

WHEN WOMEN WIN

EMILY'S LIST AND THE RISE
OF WOMEN IN AMERICAN POLITICS

Ellen R. Malcolm
with Craig Unger

Mariner Books
Houghton Mifflin Harcourt
BOSTON   NEW YORK

First Mariner Books edition 2017

www.hmhco.com

Library of Congress Cataloging-in-Publication Data
Names: Malcolm, Ellen, author. | Unger, Craig, author.
Title: When women win : EMILY's List and the rise of women in American politics /
Ellen R. Malcolm with Craig Unger.
Description: Boston : Houghton Mifflin Harcourt, 2016. | Includes bibliographical
references and index.
Identifiers: LCCN 2015037242 | ISBN 9780544443310 (hardcover)
ISBN 9780544443389 (ebook) | ISBN 9781328710277 (pbk.)
Subjects: LCSH: Women politicians — United States. | Women — Political activity —
United States. | United States — Politics and government — 1989–
Classification: LCC HQ1391.U5 M35 2016 | DDC 320.0820973 — dc23
LC record available at http://lccn.loc.gov/2015037242

Book design by Rachel Newborn

Printed in the United States of America
DOC 10 9 8 7 6 5 4 3 2 1

Small portions of chapter 9, "How to Beat Bubba," first appeared in
"How to Beat Bubba" by Craig Unger in the *New Republic,* October 22, 1990.

Graph on page 299 courtesy of Kate Black, Vice President of Research,
EMILY's List/Mapping Specialists, Ltd., Fitchburg, WI.

To those who dare to change the world:
the members, staff, and candidates of EMILY's List.

CONTENTS

THE LAST GLASS CEILING

ON THE AFTERNOON OF March 3, 2015, I arrived at the Washington Hilton to attend the thirtieth-anniversary gala for EMILY's List, the political network I founded in 1985. It was a jam-packed two-day celebration, and I had been looking forward to it for months.

Among the two thousand or so people attending the event, dozens of old friends had come not only from Washington, D.C., but also from Texas, Florida, New York, California, and all over the rest of the country. We had fought in the trenches together and forged powerful bonds. For three decades, we had shared a deep commitment to one simple but incredibly challenging mission: electing pro-choice Democratic women to office.

Young women today can't possibly imagine how quixotic that goal of getting women elected to political office seemed thirty years ago. True, the women's movement was well under way by the mideighties, and women had begun to join the workforce en masse. But having high-powered careers in law, medicine, business, or politics was a different matter entirely back then. It was for pioneers only.

Nowhere was that more evident than Capitol Hill. In 1985, 5 percent of the 435 elected officials in the House of Representatives were women. There were just two women among one hundred United States senators. And both female senators were Republicans

—which meant that my party, the Democrats, our most progressive political party, did not have a single woman in the Senate.

Not one.

Even worse, that had *always* been the case. In its entire history, the Democratic Party had never elected a single woman to the Senate in her own right. On several occasions, when a sitting senator had died, the powers that be had appointed his widow to fill her late husband's seat, and, at times, such women had even been elected to one or more subsequent terms. But when it came to electing a Democratic woman in her own right, that had never happened.

That number—zero—spoke for itself: our representative democracy had almost completely ignored 51.4 percent of the population. Virtually no one in either major political party took women candidates seriously. For generations, the old boys' network of both parties had kept women out of office. Those in the political establishment didn't think women could win, so they refused to contribute to their campaigns. And without any money to run serious campaigns, women candidates were doomed to defeat. It was a vicious cycle.

So, if you looked out on the floor of the House or the Senate in those days, you saw a forbidding sea of monochromatic suits— black, blue, or dark gray—worn by the men who had transformed the august legislative chambers of Congress into the world's most exclusive men's club. How was it possible to think of the United States of America as a representative democracy when we failed to elect women to Congress?

The situation was grim.

SO, BIT BY BIT, EMILY's List assembled a terrific team of talented women, and we came up with a sophisticated marketing strategy dedicated to winning political parity for women. At times, we suffered bitter defeats. But we made progress in each and every election cycle until, over time, we had built a full-service political organization that, in addition to raising funds, trained people to be campaign managers, to track local campaigns, to oversee media, to win, and to govern. We battled an increasingly extreme-right Republican Party led by ruthless political operatives, such as Karl Rove, and retrograde Tea Party candidates, one of whom talked about "legitimate rape" in a way that was stunning. Yet we emerged victorious again and again and eventually became, at one point, the biggest and most powerful political-action committee in the United States.

And now, in 2015, at the Hilton, one could see exactly how successful we had been. If there had been an awards ceremony for women in politics, this would have been our Oscars. With celebrities like Anna Gunn of *Breaking Bad,* Connie Britton of *Friday Night Lights* and *Nashville,* and Padma Lakshmi from *Top Chef,* there was, in fact, even a bit of Hollywood glamour. But the most honored stars were women who had real political clout. There were U.S. senators and representatives, cabinet officials, and governors. There were political professionals we had trained or worked with: consultants, pollsters, fund-raisers, strategists, and more. There was a new generation of up-and-coming politicians such as Ayanna Pressley, the first woman of color to be elected to the Boston City Council and the recipient of our Gabrielle Giffords Rising Star Award. And finally, there were real presidential contenders both for now and for the future.

If a time-lapse portrait could depict everyone there and our shared achievements, it would show nothing less than the saga of a great social movement. That saga is the story of the rise of women in American politics. It tells how EMILY's List went from nothing more than an annoying thorn in the side of the old boys' network of the Democratic Party to a powerful and highly valued partner that was absolutely essential to the party's success. It is the story of dozens of individual women who started out as envelope-stuffing activists, community organizers, local pols, or party officials, who never gave up and moved on, as a generation, into statewide politics, where some became governors, and into federal races, where they won seats in the House of Representatives and the United States Senate — and they did not stop there.

It is the story of Barbara Mikulski's journey from being a scrappy street-fighting community organizer in Maryland to making history as the first Democratic woman elected to the United States Senate in her own right. It is the story of a California Democratic Party volunteer named Nancy Pelosi, who hosted a fund-raiser for EMILY's List at her San Francisco home in the eighties and went on to become the Speaker of the House. It is the story of how the legendary Ann Richards grew from a marginalized liberal activist to an iconic figure to millions of American women — and a great governor of Texas. It is the story of a demure law professor named Anita Hill who spoke out about her sexual harassment and who, after daring to come forth, was subjected to a demeaning and insulting interrogation by the then all-male Senate Judiciary Committee, thereby igniting the passions of tens of millions of voters who voted dozens of women into higher office. It is the story of a generation-long war against an entrenched male political establishment and an increasingly right-wing Republican Party, and of unlikely wins — again and again and again.

And it is the story of how EMILY's List and all those who supported its goals reached the verge of achieving one success that initially was far beyond our wildest dreams.

I WAS SCHEDULED TO SPEAK at dinnertime. Not having given a major speech in several years, I was a bit nervous, and I arrived a few hours early to practice with the teleprompter. I rehearsed and then went upstairs to chat with several old friends who were in from out of town.

Five years earlier, I had turned over the reins of EMILY's List to Stephanie Schriock, our new president. Not being responsible for this event's scheduling glitches or anything else that might misfire, as I had been for so many years, was a real relief to me, and it meant I had some time to relax with my friends. Many of us had been separated for years by geography and circumstance, but the camaraderie instantly returned.

Then, at 6:00, I was summoned downstairs for a photo shoot with Stephanie. A few minutes later, with the Secret Service leading the way, Stephanie and I were whisked to a small but smartly decorated greenroom-like reception area. By the time we got there, the Secret Service had already made a sweep of the room. Then, at about 6:30, one of the staffers said, "She's on her way."

Fifteen minutes later, Hillary Clinton, our guest of honor for the evening and the winner of our We Are EMILY Award, arrived.

I had first met Hillary more than twenty years earlier when she was campaigning with her husband during the 1992 presidential race. For me, that year was not just the year the Clintons took the White House; it was also the Year of the Woman. In the wake of Anita Hill's explosive testimony that Supreme Court nominee Clarence Thomas had sexually harassed her, scores of women candidates were swept into office for the first time. I met Hillary at various events related to those elections.

Over the years, I grew to admire Hillary enormously, and we developed a warm, collegial friendship. Both she and Bill Clinton spoke at EMILY's List on several occasions, and, during Bill Clinton's presidency, I was invited to the White House several times as their guest.

On this occasion, it was no secret that Hillary herself was running for president in 2016 — even though she had yet to officially announce her candidacy. As it happened, however, on this particular day, the media effectively started Hillary's campaign without her, breathlessly reporting that she had used a personal e-mail account instead of the government's while she was secretary of state. There was no evidence whatsoever that Hillary had violated any regulations. Nevertheless, the press, as if to make sure Hillary's campaign was launched *not* on her terms, rallied around yet another faux scandal in the tradition of Whitewater, Travelgate, and others that had plagued Bill Clinton's presidency. Having been through this many times before, however, Hillary was unfazed.

A FEW MINUTES LATER, I entered the Hilton's vast ballroom, where 1,700 people were listening to Barbara Mikulski's remarks. It was a bittersweet moment. In addition to making history as the first Democratic woman senator elected in her own right, Barbara had spent years mentoring incoming women colleagues, telling them repeatedly it was time to "put on your lipstick, square your shoulders, suit up, and let's fight." At seventy-eight, after five terms in the Senate, she had had a spectacular career.

But the day before, Barbara had announced, in her own inimitable fashion, that she would not run for a sixth term and would step down as the longest-serving woman in the history of the United States Congress. As she put it, her decision came down to one simple question: "Do I spend my time raising money, or do I spend my time raising hell?"

Barbara had decided to do the latter. But she also made it clear that it would be foolhardy to dismiss her during the two years that remained in her term. "I'm gonna be around," she said. "I'm Senator Barb."

When the applause died down, I went up to Barbara, who was still on the stage. No one could have been a more appropriate choice to introduce me. After all, Barbara's historic victory was achieved during EMILY's first real election cycle, and her retirement marked the end of an era.

Now it was time for me to tell the assembled audience the story of EMILY's List. Back when Barbara first ran, I said, the political establishment didn't believe women could win, so they poured money into the races of male incumbents. Meanwhile, serious candidates like Barbara did everything they could to scrape together a few dollars and cents. But it was never enough.

The "Founding Mothers," as we dubbed the group of twenty-one women involved in EMILY's List from the start, had an idea.* We thought if we raised early money for our women, they could put together real campaigns and convince the establishment to hop on board. To that end, we created a donor network of men and women who would contribute to the candidates we recommended. We were like a Kickstarter for pro-choice Democratic women candidates.

It was a great strategy, but we needed lots of members. So, in April 1985, I invited about twenty-five friends to come to my basement with their Rolodexes and send letters to their friends. We asked for contributions so that EMILY's List could identify and assist good candidates. We asked members to contribute to two or three candidates whom we recommended during each election cycle. And we asked them to reach out to their friends and ask them to join EMILY's List.

In the end, our efforts worked so well that we grew from twenty-five people in my basement to three million people today. Thanks to Stephanie Schriock's new leadership, we continued to make history over and over again. We helped Nydia Velázquez and Lucille Roybal-Allard become, respectively, the first Puerto Rican and Mexican American woman elected to Congress. We helped Carol Moseley Braun become the first African American woman elected to the Senate, and Mazie Hirono the first Asian American woman elected to the Senate. We helped Jeanne Shaheen become the first woman in the history of our country to be elected both governor and senator.

And along the way, we won some of the toughest political battles in America. When former governor Tommy Thompson thought he would waltz into the Senate from Wisconsin, EMILY's List mobilized all its resources. As a result, Tammy Baldwin became the first

---

* The Founding Mothers included me, Marie Bass, Donna Brazile, Ranny Cooper, Betsy Crone, Irene Crowe, Kathleen Currie, Gail Harmon, Nikki Heidepriem, Joanne Howes, Millie Jeffrey, Kristina Kiehl, Ann Kolker, Debbie Landau, Judy Lichtman, Mimi Mager, Joan McLean, Jane McMichael, Cat Scheibner, Lael Stegall, and Mary Ann Stein.

openly gay member of the United States Senate. When Missouri's Todd Akin thought he'd bring his extreme religious views to the Senate to tell women how to run their lives, Claire McCaskill and EMILY's List intervened. As a result, McCaskill was reelected to the Senate and is fighting hard to protect women from sexual violence. And when Scott Brown from Massachusetts decided he was going to win a six-year term in the Senate, we jumped in again. As a result, Elizabeth Warren became a true progressive leader in the Senate. All told, we have helped elect nineteen Democratic women to the Senate.

In 1988, when we supported our first House candidates, there were only twelve Democratic women in the House. Since that time, we have helped elect more than one hundred Democratic women to the House. Moreover, we have helped elect more than seven hundred to state and local offices. By 2015, we had come such a long way — but we were not done yet.

BACK WHEN WE STARTED EMILY's List, there was one issue we didn't dare tackle. Being pragmatic and highly strategic was always important to us, so we focused on accomplishing what was feasible. And in the eighties, a woman becoming president was simply not in the cards. The political establishment had made it so difficult for women to win seats in the House, much less the Senate, that women couldn't possibly acquire the experience and credentials necessary to be taken seriously on the presidential level. In addition, voters didn't know what to expect from women candidates. How could they? There were so few women in office that women politicians were thought of as novelties.

But that is no longer the case. After thirty years, there is an incredibly strong bench of women with convincing credentials — senators, governors, and cabinet officials. It includes women like Sen. Elizabeth Warren (D-MA); Sen. Kirsten Gillibrand (D-NY); Sen. Amy Klobuchar (D-MN); Gov. Maggie Hassan (D-NH); Gov. Gina Raimondo (D-RI), former secretary of homeland security; Janet

Napolitano, who served as both attorney general and governor of Arizona and is presently president of the University of California; Kathleen Sebelius, who served as secretary of health and human services from 2009 to 2014 and was governor of Kansas before that; and California attorney general Kamala Harris — any one of whom might well end up on a Democratic presidential ticket someday.

And, of course, there is someone else. In fact, of all the potential presidential candidates in 2016 in both parties, the contender with the most experience and the best credentials is a former senator and former secretary of state who has already spent eight years in the White House.

"LADIES AND GENTLEMEN," I said during my speech. "I for one believe we have a little more work to do at the top of the ballot." There was a bit of laughter, and I paused for a moment before continuing. "In 2008, Hillary Clinton put eighteen million cracks in the glass ceiling. In 2016, it's time to shatter that glass ceiling and put a woman into the White House."

I looked down at the table in front of the podium to my right, and Hillary was smiling broadly, laughing, her eyes widening as all those around her — the entire room, nearly two thousand people — leaped to their feet and cheered.

"It's time to finish the job," I continued. "Are you ready for that? Do you want to elect Hillary Clinton in 2016?" With that the hall was filled with applause, and the standing ovation continued for thirty seconds. "Hillary, you heard us. Just give the word and we'll be right at your side. We're EMILY's List, and we're ready to fight for you and win!"

WHEN WOMEN WIN

One

A POLITICAL EDUCATION

I WAS AN UNLIKELY political activist. I grew up in the fifties and sixties in Montclair, New Jersey, an upper-middle-class suburb outside of New York. When I was eight months old, my father died of cancer and my mother, Barbara, became a twenty-four-year-old widow. Three years later, Mom remarried and left her job at IBM to stay home and raise her children. Her decision to quit work was never in doubt. That's what women did in the fifties — if the family could afford it.

When I entered Hollins College, an all-women's school in Roanoke, Virginia, in 1965, I was an eighteen-year-old preppie who was essentially apolitical. This was an era when men's schools and women's schools were more common than they are today, and it did not even occur to me to go to a coed school. I didn't even really know the difference between Democrats and Republicans, liberals and conservatives. I'd never heard of Vietnam, much less realized we were at war there, and I didn't even know that hundreds of thousands of Americans were protesting.

But, in 1968, at the urging of a friend, I went to Philadelphia to work for Eugene McCarthy, the antiwar senator from Minnesota, during the Democratic presidential primary in Pennsylvania. I knocked on doors, handed out literature, and talked to people about the issues. McCarthy won 71 percent of the vote in Pennsylvania. I had just turned twenty-one and was now eager to vote in my first presidential election.

It was 1968 in America. All across the country, the counterculture of the sixties was ascendant. A generation of antiwar protesters and long-haired hippies were replacing buttoned-down, crew-cut frat boys and sweater-and-pearls sorority sisters. Bob Dylan, the Beatles, and the Rolling Stones were on the airwaves. On the other side of the globe, the North Vietnamese launched the Tet Offensive against South Vietnam. U.S. campuses were in an uproar. On March 31, President Lyndon Johnson announced he would not seek reelection.

Four days later, on April 4, 1968, I was crossing the Hollins campus when I heard that Rev. Dr. Martin Luther King Jr. had been assassinated. Hundreds of thousands of people rioted in New York, Washington, Chicago, Los Angeles, and dozens of other cities. On June 6, Bobby Kennedy was assassinated in Los Angeles. There were countless demonstrations all over the country.

My political innocence was over. Both Montclair and my family were Republican to the core, but I headed off in a very different direction. Too much of what was happening in the sixties was close to home — literally. Just eleven miles from Montclair, Newark was the epicenter of the most violent racial upheavals of the time. The year before, six days of rioting, looting, and violence left 26 people dead, more than 700 injured, and 1,500 arrested — not to mention millions of dollars in damages. In the aftermath of the King assassination, civil unrest spread to 125 cities.

To affluent suburbanites like me, all this was shocking. By that summer, I was fully committed to fighting poverty and racism, as well as the war. I believed that job training could help the unemployed, so my mother found me a volunteer job at the Manpower Development program in Newark. There I was, a nice, young, MG-driving white girl from Hollins, whose mother was urging her to join the Junior League, working on Broad Street in Newark — not far from the riot-torn ghetto. It was an eye-opening experience.

In August, the McCarthy campaign sent out word that staffers and volunteers should stay away from the Democratic National Convention in Chicago, so I watched it on TV from the comfort of home. In the aftermath of the MLK and RFK assassinations, Chicago was the apex of counterculture protest. For five days, thousands of Chicago police fought demonstrators in the streets, while nearby, in the International Amphitheater, the Democratic Party selected Vice President Hubert Humphrey as its presidential nominee.

By the time I returned to campus for my senior year, I had become Hollins's version of a campus activist. Granted, I didn't build bombs or take over the administration building. But I had changed. I went to Charlottesville to hear blues singer Janis Joplin give us "Another Piece of My Heart." I went to civil rights meetings. And, in my own decorous way, I did something audacious: I invited the college president to the campus dining hall as part of a campaign to allow Hollins's students to wear pants.

As a measure of exactly how radical I was, I took the defiantly militant step of wearing pants to the meeting. Lo and behold, soon enough the rules were changed. It wasn't exactly rabble-rousing, but it was my first taste of political success, and I loved it.

IN THE SUMMER OF 1970, I moved to Washington, D.C. I had learned about a new nonpartisan, grassroots "citizens' lobby" called Common Cause that focused on campaign-finance reform, election reform, accountability, and the media. I arrived when the organization was just six months old. Our first goal was to end the Vietnam War.

Common Cause took a less attention-getting but more pragmatic tack than those used by most antiwar groups: its goal was to cut off the federal funds that allowed the war to continue. My job was to oversee a small army of volunteers to mobilize public pressure on senators and representatives to pass our agenda. We didn't have the money that the special interests had, but we had created a new kind of organization as a counterweight: a citizens' lobby that harnessed the voices of ordinary Americans. I focused on organizing volunteer lobbying groups in congressional districts and on keeping volunteers informed and excited so they would work successfully. In addition to our antiwar efforts, we initiated campaign-finance-reform legislation to limit how much money individuals and organizations could give to candidates, to make those contributions public, and to establish an independent organization to oversee campaign financing.

Campaign-finance reform wasn't the kind of issue that captured the imagination of the American people. But it soon became clear that our work was part of something much bigger. In 1971, Common Cause sued the Democratic and Republican Parties, asserting that both parties were ignoring the Federal Campaign Practices Act of 1925. In response, Congress quickly passed the Federal Election Campaign Act to increase disclosure of campaign contributions. The law went into effect on April 7, 1972, just as the new presidential season got under way.

The names of the Republican donors soon turned out to have enormous historic value. The reason? Three months later, on June 17, 1972, five men paid by Nixon's Committee to Reelect the President (aptly shortened to CREEP) broke into and entered the headquarters of the Democratic National Committee in the Watergate complex in Washington. Over the next two years, as the Watergate scandal unraveled on national TV, the names of those donors became essential to following the money—which was the key to getting to the bottom of the entire scandal. And the reason those names became public was that Common Cause sued CREEP, thereby forcing it to reveal who had contributed millions of dollars—much of it cash literally stuffed into suitcases and satchels.

As a result, I had a front-row seat to one of the greatest political spectacles of the century: a psychodrama about paranoia and power starring Richard Milhous Nixon that had the entire nation glued to its TVs. A generation before anyone had even heard of binge-watching, this was reality TV.

It all came to a head on August 8, 1974, when, rather than face impeachment by the House of Representatives and near-certain conviction by the Senate, Nixon became the first president of the United States to resign from office. To this day, I can see Richard Nixon stopping at the door of his helicopter, awkwardly waving good-bye to his presidency and to his reputation in history.

Shortly after Nixon's departure, Congress passed a sweeping campaign finance-reform bill designed to minimize candidates' reliance on huge donations from special interests. Not long after that, Congress cut funding to South Vietnam from a proposed $1.26 billion to $700 million. A year later, in 1975, the United States withdrew all troops from South Vietnam. Saigon fell. I was proud of what we had done: the funding cuts that Common Cause had lobbied for had finally brought an end to the Vietnam War.

My political education had begun. I had learned how electoral politics work, how bills are written and become law, and the nuts and bolts of campaigning. I had learned how political campaigns are funded, how lobbyists effectively buy access to incumbent members of Congress, and how that access leads to legislation that serves the lobbyists' interests. All of this knowledge would come in handy in the future.

CIVIL RIGHTS AND THE antiwar movements weren't the only social crusades sweeping America. In 1960, just forty years after women finally won the right to vote in the United States, the widespread availability of the birth control pill for the first time gave women the power to separate sexual activity from procreation, thereby allowing them to postpone having children so that they might pursue careers that had long been out of reach.

Soon, a new wave of feminists took on issues such as reproductive rights, sexuality, discrimination in the workplace, and various other inequities. In 1963, *The Feminine Mystique,* Betty Friedan's landmark best seller, described what she called "the problem that has no name" — namely, the plight of women who, despite living in material comfort and being married with children, were unhappy. In large measure, that was because they were treated as second-class citizens in both their personal and professional lives.

Even though feminists were often lampooned as bra-burning "women's libbers," the movement continued to grow.* One milestone followed another. In 1966, the National Organization for Women (NOW) was founded, with Friedan as its president. Journalist Gloria Steinem became a major feminist voice. In 1968, the word *sexism* was coined. The next year, NARAL, the National Association for the Repeal of Abortion Laws, was launched. In 1970, Kate Millett's *Sexual Politics* and Germaine Greer's *The Female Eunuch* became must-reads. In 1972, Shirley Chisholm, the first African American woman elected to Congress, became the first major-party black candidate for president. Through it all, there was "Battling Bella" Abzug, the brassy and exuberant New York congresswoman known for her famously floppy hats and a voice Norman Mailer said "could boil the fat off a taxicab driver's neck." Asserting that "this woman's place is in the House — the House of Representatives," Abzug gave a feisty feminist cast to the few women who were entering electoral politics.

And last, but hardly least, in 1973, the landmark Supreme Court decision *Roe v. Wade* legalized abortion, creating a litmus-test issue that forever redefined the battleground of reproductive rights.

By that time, Gloria Steinem, Letty Cottin Pogrebin, and others had launched *Ms.* magazine, which quickly became a feminist bible chronicling the ascent of feminism not just as a social move-

---

* It's worth noting that bra burning never really happened. In 1968, *New York Post* reporter Lindsy Van Gelder wrote about a feminist protest at the Miss America Pageant at which bras and girdles were going to be tossed into a bonfire. The bonfire never took place, but Van Gelder compared it to the burning of draft cards by antiwar protesters — and a myth was born.

ment but as a new consciousness. As Jane O'Reilly put it in the
magazine's introductory issue, the "click" was the "moment of truth
. . . the shock of recognition," the sudden awareness that what was
widely believed to be the natural order of things in gender relations
was in fact shockingly out of whack. It was the realization that men
and women often used very different languages in talking to each
other, and the awakening to the reality that millions of enormously
bright, accomplished women were frustrated and unfulfilled, rel-
egated to servile roles where they were paid little or no attention,
and imprisoned by rigid and limiting expectations as housewives.

IN THE CONTEXT OF all that, exactly where did I stand?

I was certainly aware of Bella Abzug, Gloria Steinem, and *Ms.*
magazine, but the women's movement was largely foreign to me. I
didn't even think of myself as a feminist.

In Montclair, my mother's path to employment as a secretary
required one essential skill: typing. That's how women got into
the corporate world back then. Many years later, I realized that my
mother's skills were equal to those of many executive men. As president of the Junior League of Montclair, as chair of the local Essex
County branch of the United Way, and through her other work, she
managed budgets, gave speeches, brought meetings to order, and
won elections. Yet to Mom, it was a given that men were the breadwinners and women took care of the kids. I was brought up to believe that women could be secretaries, teachers, or nurses, but there
were limits as to how far they could go. Before college, it hadn't
even occurred to me that women could be professionals. But all that
was beginning to change.

When they did enter the professional world, however, even the
most highly credentialed women encountered various forms of
sexism. Many years later, I discovered that this was the case for a
twenty-one-year-old senior at Wellesley College who went into
Cambridge, Massachusetts, one day to take the LSATs so she could
apply to law school. Once she got there, Hillary Rodham encoun-

tered several young men who didn't like the idea that women were competing with them. This was a period during which a law-school student deferment meant exemption from the Vietnam War draft, and one of them told her, "If you take a spot, I could end up going to Vietnam. You'll be responsible for me dying."

"It sounds ridiculous now," Hillary told me, "but there wasn't a hint of humor in it."

After being accepted by both Harvard and Yale Law Schools, Hillary returned to Cambridge to attend a Harvard cocktail party where a friend introduced her to a well-known law professor. "This is Hillary Rodham," the friend said. "She's trying to decide whether to come here next year or sign up with our closest competitor."

Without missing a beat, the Harvard professor looked down at Hillary dismissively. "First," he said, "we don't have any close competitors. And second, we don't need any more women at Harvard."

That made Hillary's decision simple: she went to Yale.

Hillary, like so many women of her generation, continued to encounter archaic attitudes about gender in even the most progressive precincts of the law. In 1974, as a twenty-six-year-old graduate of Yale Law School, Hillary was the only woman lawyer on the House Judiciary Committee staff during Richard Nixon's impeachment hearings, having been hired by John Doar, the chief counsel for the committee and an iconic figure in the American civil rights movement.* At the time, Hillary was dating Bill Clinton, who had gone back to Arkansas to run for Congress, but, like Doar himself, she often worked late into the night. "I was there at roughly midnight one night, doing my work," Hillary said. "And he was wandering around, seeing what work people were doing. By then, he knew that I was dating Bill. And here this avatar of the civil rights movement comes in and says, 'Can I ask you a question?'"

---

* When David Hume Kennerly, the Pulitzer Prize–winning photographer, photographed Doar during this period, Kennerly wanted to crop out the pesky young woman at Doar's side. It was Hillary.

"Sure, Mr. Doar," Hillary replied.

"Well, why are you here while your boyfriend is running for Congress? You're not supporting him."

Doar went on to say that he was delighted that Hillary was there working late into the night. But there it was: even when highly accomplished women were working in the most progressive circles imaginable, a man's work *always* took precedence over his mate's.

That needed to change, too.

Two

A MOVEMENT BEGINS

AFTER FOUR AND A HALF YEARS, I left Common Cause. It took me about another year of part-time work to find what I wanted: a job as press secretary for the National Women's Political Caucus (NWPC). The caucus was the first national organization dedicated exclusively to increasing women's participation in the entire political sphere — as elected and appointed officials, as delegates to Republican and Democratic conventions, as federal and state judges, and as lobbyists, voters, and campaign organizers. Before the 1976 presidential race, the caucus met with Jimmy Carter and won support for proposals for appointing women to the cabinet and the Supreme Court. Thanks largely to the efforts of NWPC chairwoman Mildred "Millie" Jeffrey and NWPC Democratic Task Force chairwoman Joanne Howes, the party changed its rules to mandate "equal division" of delegates by gender at future Democratic conventions. That last groundbreaking decision meant that, starting in 1980, half the delegates at the Democratic National Convention would be women.

It was in this context that I began working for the NWPC, eventually as a press secretary, in pursuit of its extraordinarily ambitious primary objective: the ratification of a proposed amendment to the Constitution designed to guarantee equal rights for women — the Equal Rights Amendment. The mere fact that an amendment to the Constitution was necessary to guarantee equal rights to women was astonishing—but it was, and it had been since time immemorial.

Gradually, I learned that the women's movement was a historic undertaking in its infancy. Like the civil rights movement and the antiwar movement, this was a struggle that would take decades. And, like those other social movements, the greatest breakthroughs usually came about as the result of hard work and ingenious strategic maneuvering.

Indeed, as scholars later revealed, savvy activists went to extraordinary lengths to hoodwink conservative legislators into serving the feminist cause. The Civil Rights Act of 1964, one of the most important pieces of legislation of the century, was a case in point. The impetus behind the bill, of course, was to ban racial discrimination as a means of honoring the memory of President Kennedy in the aftermath of his assassination. But while the bill was still being debated, Rep. Howard W. Smith (D-VA), the chairman of the Rules Committee, a staunch conservative and a steadfast opponent of civil rights legislation, proposed a one-word change to Title VII of the bill, the section that was initially written to prohibit discrimination against any individual with respect to compensation, terms, conditions, or privileges of employment because of such individual's race, color, religion, or national origin. Smith suggested that the word *sex* be added to the bill to "prevent discrimination against another minority group, the women."

From today's vantage point, Smith might appear to have been motivated by noble and progressive concerns. But at the time, his modification of the bill was widely seen as an attempt to kill the entire bill. That's because in many circles the idea of equality for women was perceived as so ludicrous that no one could possibly vote for it. In fact, when Smith introduced the wording change, laughter broke out on the House floor.

This chapter of feminist history is exceptionally counterintuitive in that many women leaders, from Eleanor Roosevelt on down, actually *opposed* the Equal Rights Amendment and similar measures, because they feared such laws would jeopardize hard-won legislation that protected women in the workplace. Conversely, the biggest supporters of the ERA tended to be conservative anti-union businessmen in the South, because owners of textile mills, which employed large numbers of women, stood to benefit from fewer workplace regulations.

These strange crosscurrents brought together such unlikely bedfellows as Howard Smith and Rep. Martha Griffiths (D-MI). Smith, on the one hand, a segregationist who consistently backed legislation benefiting textile mills in his state, had supported the ERA for more than twenty years. Griffiths, on the other hand, came to the issue with straightforward feminist motives—and extraordinary political talents. A lawyer and judge before joining Congress, Griffiths was the first woman to serve on the powerful House Ways and Means Committee. Sometimes known as the mother of the Equal Rights Amendment, she deployed, as *The Guardian* put it, an "implacable determination, a lawyer's grasp of procedural niceties, and a tongue like a blacksmith's rasp."

During her ten-term career, Griffiths uncovered countless astonishing inequities in federal government policies and fought to rectify them again and again. Social Security, she discovered, would pay benefits to a dead man's children—but not to a woman's. Similarly, if a man survived his wife, he was exempt from paying taxes on her estate, but widows were not exempt when their husbands died. There were analogous inequities when it came to pension eligibility. Nor was Griffiths shy when it came to delivering withering attacks on her adversaries—as when she discovered that the airlines fired stewardesses if they got married, and hired only women who were young, attractive, and single. "What are you running," she wrote in a letter to one airline's chief executive, "an airline or a whorehouse?"

When it came to the Civil Rights Act, Griffith saw a historic opportunity and had no qualms about seizing it by forging an alliance with a segregationist congressman from the South. "She thought once the word *sex* was in, nobody would take it out," said Judy Lichtman, a civil rights lawyer who came to the Women's Legal Defense Fund in 1974 and helped pass related legislation over the next forty years. "So, she intentionally made alliances with these Neanderthals who wanted to undermine the Civil Rights Act."

Rather than introduce the amendment herself, Griffith decided to let Howard Smith carry the ball. From his point of view, it was a no-lose proposition. If the word *sex* torpedoed the historic bill, Smith would look like a hero to his racist constituents. If it passed, that would endear him to the textile-mill owners in Virginia.

When Smith brought the bill before the House, as Louis Menand pointed out in *The New Yorker* in 2014, he did so as if the whole issue were a joke, reading a letter from a constituent calling for the government to attend to the "right" of unmarried women to find a husband. That elicited laughter. Then, liberal supporters of the Civil Rights Act spoke out against the "sex" addition because they regarded it "as either a prank intended to expose the limits of liberal egalitarianism or a poison pill that would make the bill more difficult to pass." Emanuel Celler (D-NY), the chairman of the House Judiciary Committee and a strong proponent of the Civil Rights Act, was less than enthusiastic about including "sex" as a protected category. He knew all about equal rights for women, he claimed, because, after forty-nine years of marriage, he always got the last two words: "Yes, dear."

The jokes continued unabated, demonstrating that even in these august chambers women were seen as nothing more than second-class citizens. At last, Griffith finally took the floor and prevailed, and the addition of the word *sex* to Title VII of the Civil Rights Act became a legislative landmark in the history of women's rights.

IMPLEMENTING THIS NEW BAN against gender discrimination, however, led to yet more battles. Title VII was one of the most demanding components of the Civil Rights Act, and it necessitated the creation of the Equal Employment Opportunity Commission, whose chairman, Franklin D. Roosevelt Jr., argued that the prohibition of sex discrimination was not to be taken as seriously as the outlawing of racial discrimination.

Over time, however, the law proved to be a momentous break-through, because it provided a powerful basis for litigation against employers who discriminated against women. "It had an enormous impact," said Judy Lichtman. "For the first time, it gave women the legal right not to be discriminated against in the workplace. As a result, employers changed their behavior. Women who were discriminated against brought lawsuits. One administration after another began to promulgate laws that prohibited discrimination. Classified ads in newspapers for employment could no longer be segregated by sex. Women's economic opportunities were vastly changed and enhanced by Title VII. It was absolutely huge."

VITAL AS TITLE VII WAS, it focused narrowly on employment, and the Equal Rights Amendment was the logical next step to provide a far broader range of protection to women's rights. Both houses of Congress passed the amendment in 1972, but constitutional amendments have to be ratified by three-quarters of the state legislatures in the United States. Getting thirty-eight state legislatures to sign off would not be easy. When the National Women's Political Caucus convinced President Carter to support the ERA, the women's movement had begun to move from engaging in mere policy discussions to joining forces with political power.

At the time, I had just turned thirty and was still finding my professional sea legs. But I felt inspired thanks to the chair of the NWPC, a diminutive, disheveled, and dynamic powerhouse of a woman who became my role model.

In an era in which antiwar leaders, counterculture gurus, and rock stars vied for their fifteen minutes of fame, Millie Jeffrey, then in her mid-sixties, was the real deal, a creature of history who was deeply committed to building institutions that would endure, to creating lasting social change that would stand for generations. Just four feet eleven inches tall, Millie had decades of experience building bridges between the civil rights movement and the labor movement. Above all, she was committed to equal rights for workers, for minorities, and for women.*

In the thirties and forties, Millie had been a labor organizer with the Amalgamated Clothing Workers of America and the United Auto Workers, where she won access to its inner chambers thanks to her close relationship with its president, Walter Reuther, the great hero of the American labor movement. In the fifties and sixties, she had become an activist in the civil rights movement, registering voters in Mississippi and marching with Rev. Dr. Martin Luther King Jr. First in John F. Kennedy's 1960 presidential campaign, and later in Bobby Kennedy's ill-fated 1968 presidential campaign, which ended with his assassination, she was, as always, a builder of bridges. It was Millie who convinced Reuther to join the civil rights marches in the South, thereby cementing a vital alliance between the civil rights movement and organized labor. Likewise, it was Millie who persuaded Michigan civil rights leaders to pose with Sen. John F. Kennedy on the steps of his Georgetown home, thus orchestrating a political triumph for the fledgling presidential candidate.

When it came to the NWPC, Millie's diminutive physical stature belied the immense impact she had there, imbuing our work with a moral force and a sense of history. As for work ethic, it seemed that Millie was *always* in the office. As I later found out, the reason for her constant presence, sadly, was that Millie had only a meager union pension, and she couldn't afford an apartment. She actually lived at the office, sleeping on the floor. (As soon as the staff realized that, Millie became a frequent overnight guest in many a household.)

THE MOST IMPORTANT WAY in which the NWPC differed from EMILY's List today was quite fundamental: it was bipartisan. That

---

* In 2000, President Bill Clinton awarded Millie the Presidential Medal of Freedom, the highest civilian award in the United States.

was possible because, in those long-ago days before the Reagan era, many moderate Republicans — who have since gone the way of the dinosaur — were pro-choice and for the ERA, positions that would not fly in today's ultraconservative GOP. The caucus was comprised of women who were not merely paying lip service to feminist ideals. Margaret Heckler, the longtime "Rockefeller Republican" congresswoman from Massachusetts, was a strong supporter of equality for women in Social Security and tax laws, and a powerful proponent and cosponsor of the ERA. Likewise, Republican senator Nancy Kassebaum (R-KS) championed issues such as reproductive health, child care, and the ERA. Most importantly, both parties had task forces working together on the ERA, appointments, and other issues.

Such bipartisanship notwithstanding, by the late seventies, the ERA had been losing momentum. In 1973, just after the amendment had passed both houses of Congress, thirty out of the thirty-eight states necessary had ratified it. But that same year, the passage of *Roe v. Wade* had ignited the forces of the right, and ratification slowed to a crawl. In 1974, three states more ratified the ERA; the following year, just one. In 1976, there were none. In 1977, Indiana became the thirty-fifth state to ratify, but passage was three states short. Because the seven-year time limit was running out, the caucus began a lobbying effort to extend the ratification time.

The ERA had run head-on into a right-wing constitutional lawyer and leader of the New Right named Phyllis Schlafly, whose STOP ERA campaign led the way against ratification. As author Tanya Melich has pointed out in *The Republican War Against Women,* "Schlafly herself fit the feminist model of a self-assured woman with a cause, who speaks out and organizes for what she believes regardless of traditional constraints." A high-powered political figure who spurned homemaking for a career, she demanded that other women forgo their professional ambitions and instead hew closely to the role of stay-at-home moms. "Women's lib is a total assault on the role of the American woman as wife and mother and on the family as the basic unit of society," wrote Schlafly. "Women's libbers are trying to make wives and mothers unhappy with their career, make them feel they are 'second-class citizens' and 'abject slaves.'"

Thanks to Schlafly, themes of homemaking, motherhood, and "family values" became potent weapons in the Republican arsenal for decades to come. Fear of women's liberation became a powerful force. Misogyny was celebrated. Schlafly argued — wrongly — that the ERA would eliminate gender privileges enjoyed by women, including "dependent wife" benefits under Social Security and exemption from the military draft. One Schlafly ally cautioned that "the ERA would mean the end of femininity." Others asserted that privacy rights would be overturned, that women could no longer be supported by their husbands, and that the ERA would lead to the horrifying prospect of unisex toilets.

Ultimately, Republicans equated the women's movement with the civil rights movement and rolled out talking points analogous to the race-baiting tactics of the GOP's Southern strategy. One Republican woman said, "Forced busing, forced mixing, forced hiring. Now forced women. No thank you." In North Carolina, another beseeched local officials not to "desexegrate us."

As the 1979 deadline for ratification approached, the NWPC worked with the League of Women Voters, NOW, and many other organizations to pressure Congress to extend the deadline. In July 1978, NOW organized a march of one hundred thousand supporters in Washington, and Congress responded by extending the deadline to 1982.

But the tide of conservatism was rising. Millions of southerners saw the ERA as an assault by big government on all they held dear — their families, their homes, motherhood, and traditional relationships between husbands and wives. The moderate Republican women who were in NWPC may not have known it yet, but they were an endangered species. The Reagan era was nigh. The upshot was that the ERA never passed the state legislature in key states such as Florida, Illinois, and North Carolina. In the end, it never became law.

EVEN THOUGH IT LOST the battle for the ERA, the NWPC had made significant strides in what was to be a much longer war. "It [the ERA effort] was important because it taught us how to do politics," said Betsy Crone, who handled direct mail for the caucus. "It was a process we had to go through. We were inventing this movement, step by step, bit by bit. Each project we tackled helped us reach out to women all across the country."

A new dynamic was at work, an organic process through which women were becoming politicized and were finally addressing their own interests. In 1972, Shirley Chisholm ran for president in the Democratic primaries. In 1973, *Roe v. Wade* was decided. In 1975 came the International Women's Year, followed by the United Nation's Decade for Women in 1976. Also in 1976, Rep. Barbara Jordan (D-TX) became the first African American woman to give the keynote address — and a memorable one, at that — at the Democratic National Convention.

The NWPC didn't let up. Jimmy Carter had established the Committee of 51.3 Percent as part of his promise to find qualified women appointees, but the caucus made certain he honored his commitment. Even before he took office, Rep. Barbara Mikulski (D-MD) chastised the president-elect for relying too heavily on the "old boys' network." Then, as soon as he moved into the White House, the caucus, NOW, and the American Association of University Women lobbied President Carter for cabinet and subcabinet-level appointments for women in the administration, particularly in the departments of interior, labor, and health, education, and welfare.

These strategies paid off. A week after the caucus sent a Mailgram in support of Juanita Kreps, President Carter appointed her secretary of commerce. Patricia Roberts Harris became secretary of housing and urban development and, as such, was the first African American woman to hold a cabinet post. Soon, the number of women named to appointed positions in her department increased to 49 percent. In the judiciary, Carter appointed no fewer than forty women judges to federal courts — five times as many as all his predecessors combined.

The NWPC pushed for enforcement of federal statutes prohibiting discrimination against women. As general counsel of the American Civil Liberties Union, Ruth Bader Ginsburg argued one landmark case after another on behalf of women's rights before the United States Supreme Court. We created new fund-raising networks. We organized.

At the same time, Millie Jeffrey continued to press for "equal division" of delegates at the Democratic Convention — meaning that 50 percent of them would be women. Historically, African American civil rights activists had feared that gains for women were likely to come at their expense. To head off such concerns, in 1978, Millie met with Yvonne Brathwaite Burke, the African American representative from Los Angeles. "We had to show that we were not just taking care of ourselves, that we were worrying about other people's interests," said Betsy Crone. "Yvonne Burke was key to getting African American women to see that this was a good thing for them as well."

This was political coalition building at its best. As a result, that year the Democratic Party changed its rules so that women would make up half the convention delegates at the 1980 convention and all subsequent conventions. This meant an enduring institutional change in the Democratic Party. "The 'equal division' rule had a profound impact on the party," said Joanne Howes, who was head of the NWPC Democratic Task Force and worked for Barbara Mikulski in Congress. "It gave women another entry point into politics." And it created a foundation so that women could win higher positions in years to come.

Millie had developed a cadre of deeply committed and sophisticated women activists who were acquiring genuine political skills and experience. Joanne Howes had cut her teeth working for the George McGovern campaign in 1972 before going to the NWPC and lobbying for the ERA. Betsy Crone had interned in the Senate while in college, working for Sen. Joe Tydings (D-MD) and then learning political fund-raising in the late seventies at the feet of Matt Reese, a pioneer of political consulting who was among the first to combine demographic data, polling, and computers to identify voter blocs for whom specific issues were crucial. Working with Reese, who had a number of women politicians as clients, Betsy learned firsthand just how difficult it was to do fund-raising for women politicians.

In addition to Betsy and Joanne, there was Lael Stegall, the NWPC development director; Judy Lichtman, who was executive director of the Women's Legal Defense Fund (now known as the National Partnership for Women and Families); and Marie Bass, who was the campaign director of ERAmerica, a coalition of groups working to pass the ERA.

And then there were the candidates. Mikulski herself was part of a new generation of smart, ambitious young women all over the country who were emerging as rising young stars in politics. The great-granddaughter of Polish immigrants, Mikulski had made a name for herself in the seventies as activist social worker who successfully fought plans to build a sixteen-lane highway through the Fell's Point and Canton neighborhoods of Baltimore. In 1976, she was rewarded with a landslide victory that sent her to Congress.

Similarly, Ann Richards was the beautiful, charismatic founding member of the Texas Women's Political Caucus who had been elected Travis County commissioner and was clearly destined for bigger things. In California, an up-and-coming young woman named Nancy Pelosi, the daughter of Thomas D'Alesandro Jr., a former congressman and mayor of Baltimore, had, after raising five children, given in to the political blood in her veins, and become chair of the Democratic Party in Northern California. In the House were women like Pat Schroeder (D-CO), Bella Abzug (D-NY), Shirley Chisholm (D-NY), and Martha Griffiths (D-MI).

And, lest I forget, after her stint on the House Judiciary Committee, Hillary Rodham had moved to Arkansas and married Bill Clinton. Yet while he climbed the political ladder, being elected first attorney general and then governor of Arkansas, Hillary did anything but abandon her career, cofounding Arkansas Advocates for Children and Families, becoming the first female chair of the Legal Services Corporation, in 1978, and becoming the first female partner at the Rose Law Firm, the oldest law firm west of the Mississippi.

These were just some of the women who pioneered the movement that redressed gender inequality — and they were going places.

Three

STARTING FROM ZERO

MARK HANNA, THE LEGENDARY political kingmaker who put William McKinley in the White House more than a century ago, was famous for saying, "There are two things that are important in politics. The first thing is money, and I can't remember what the second one is." There was considerable truth there, but for women in politics, money was a special problem — especially given that women had just begun entering the workplace in various professions.

As a fund-raiser for the NWPC, Betsy Crone knew all about it. "Women were not used to giving at all," she said. "And those who did, didn't give in big chunks. It was twenty-five or thirty-five dollars. Beyond that, the most frequent line you heard was, 'I'll have to ask my husband.' In those days, they didn't see it as their own money. There were almost no big checks."

One exception to the rule was the Women's Legal Defense Fund, headed by Judy Lichtman, the pioneering attorney whom I had quickly grown to admire. A few years older than me, Judy was one of only 2 women in a class of 150 to graduate from the University of Wisconsin Law School in 1965. She had begun her career as a civil rights lawyer working on school desegregation in the South, and later returned to Washington, where she worked on school desegregation for the Urban Coalition and explored issues involving race and gender for the U.S. Commission on Civil Rights. By the time I started work at the NWPC, she was executive director of the Women's Legal Defense Fund (later the National Partnership for Women and Families), and through it she played key roles in virtually every civil rights advance for women over the next forty years.* Few people were more knowledgeable about affirmative action and its impact on blacks and women.

Starting in the mid-seventies, Judy would get an anonymous donation to the Women's Legal Defense Fund of $5,000 every once in a while. Sometimes the checks were for $10,000. And they all came from the Boston Safe Deposit and Trust. Every time a new check arrived, Judy was delighted, of course — but puzzled. This kind of thing just did not happen. It was like John Beresford Tipton on that 1950s TV series *The Millionaire*. Dying to find out who her benefactor was, she called the Boston bank. When it refused to divulge her patron's identity, Judy asked whether the donor was a man or a woman.

"When was the last time a man contributed $10,000?" came the response.

Later, at a NWPC staff meeting, Lael reported that a new $5,000 check had arrived from the same anonymous donor. I was in the meeting, trying desperately not to show my embarrassment.

What no one knew was that I had a secret. When I was growing up in Montclair, mine appeared to be just another a typical upper-middle-class family with a station wagon and two golden retrievers. But that wasn't the whole story. I didn't broadcast it, but my great-grandfather A. Ward Ford was one of the founders of IBM. Because my father had died when I was a child, I inherited money when I

---

* Judy played a key role in the passage of laws such as the Pregnancy Discrimination Act of 1978, the Civil Rights Act of 1991, the Health Insurance Portability and Accountability Act of 1996, and, most notably, the Family and Medical Leave Act of 1993.

turned twenty-one. I was constantly on the lookout for ways to put that money to use effectively and discreetly.

One day, in 1979, I was having lunch with Lael Stegall, the caucus's development director, at a small bistro on Capitol Hill. A passionate adventurer and activist, Lael had worked on a lobster boat in the North Atlantic, had gone dogsledding in Maine, was a Peace Corps volunteer in Turkey — and had become my friend. But now, she told me, she was ready to leave the caucus and was in the process of developing a new job. Thanks to her fund-raising efforts at the caucus, she had encountered all sorts of wealthy women and wanted to help them learn how to deal with their money and function effectively as philanthropists.

She didn't realize it, but she was talking about women like me.

By this time, it was impossible for me to remain quiet. So, I leaned across the table and confessed that I was the anonymous donor from the Boston bank.

Lael was amazed, so I explained my dilemma. I was working in nonprofit institutions, and I wanted to build my career without being labeled as a donor. I wanted to give away even more money, but I needed to know more about potential recipients so I could make sensible business decisions, and I couldn't figure out how to do that while still maintaining my anonymity.

"I could be much more effective and more generous if I knew where best to put the money. Could you help?"

As a result of that conversation, in 1980 I founded the Windom Fund, with Lael as executive director, working out of 2000 P Street. I swore Lael to secrecy about my role. It was perfect. I could now continue to fund the foundation secretly, secure in the knowledge that my philanthropy was based on sound information and well-thought-out strategy.

IN JULY 1979, Millie Jeffrey was defeated for reelection as chair of the caucus, and, after several months of struggling with her replacement, it was time for me, like Lael, to move on. I did pro bono work for a few months and was then hired as press secretary to Esther Peterson, the special assistant for consumer affairs in the Carter White House.

Like Millie, Esther had been a powerful influence on me. A commanding figure in the American labor movement, after working as an organizer for the American Federation of Teachers in the thirties, she became the first lobbyist for the National Labor Relations Board, in 1944, and the first woman lobbyist with the AFL-CIO, in 1957. Back then, Esther lacked seniority, which put her at the bottom of the totem pole, but even that, she told me, led to unexpected perks. "They gave me the [congressional] members with the least influence," she said. "One of them was a handsome man whose wealthy father supposedly bought his congressional seat." His name was Jack Kennedy, and he and Esther soon became fast friends.

When Kennedy became president, he made Esther the highest-ranking woman in his administration, naming her assistant secretary of labor and director of the United States Women's Bureau. By the time she hired me, she had become Jimmy Carter's director of consumer affairs. Among Esther's lasting accomplishments was getting foods labeled with their nutritional value, and getting products to be priced per unit so that consumers could easily compare prices and determine which product was the best value.

Like Millie, Esther had a strong affinity with organized labor. Raised as a Mormon in Provo, Utah, she had moved to Cambridge, Massachusetts, with her husband, where she taught school. In the thirties, she took the bus out to Lowell, Massachusetts, where she led striking mill workers in song on the picket lines. After she moved to Washington, Esther's ties to labor helped her become the first lobbyist for the National Labor Relations Board.

During the 1980 presidential campaign, Esther and I drove through Lowell, and she told me about the women textile workers on strike at the Lowell mills in the 1830s, when girls as young as ten years old worked more than seventy hours per week in incredibly hot and noisy conditions. The strikes led to the creation of the Lowell Female Labor Reform Association, the first union of women workers in the United States, and that in turn led to years of organizing to reduce the number of hours in the workday and to improve pay and working conditions. Like Eleanor Roosevelt, whom she knew and worked with, Esther opposed the Equal Rights Amendment for many years because she feared it would jeopardize hard-won victories for working women.

My mother had taught me that it was important to give back, but I had no real political values until Millie Jeffrey and Esther Peterson showed me where I really belonged. Whether it was fighting for rights of working men and women, minorities, or consumers, they battled for progressive values of fairness, equality, and opportunity. They showed me that social activism means bringing people together even if you don't always agree with them. I was always struck by their quiet determination and by the way their lives reflected their ideals.

Most significantly, I learned from them that social change takes constant and consistent commitment, and that sometimes we make incremental progress, if that, until events converge and we can take a giant leap forward. They showed me that there were other women who had come before me, that what I was starting to do was part of a movement, part of history, and that it had to be done with an eye toward making history over a multigenerational long struggle, because social change can be a very slow business. Given that we were about to witness the birth of the Reagan era, I soon came to realize that we would need a great deal of patience.

ON JULY 14, 1980, the Republican National Convention nominated Ronald Reagan and George H. W. Bush as its party's respective presidential and vice presidential candidates. The Reagan Revolution was born, and with it the "family values" promulgated by

Phyllis Schlafly — code words of the day signifying support for traditional marriage and sexual abstinence outside of marriage, as well as opposition to legalized abortion, to feminism, to homosexuality, and more. These trends were very much on the ascent.

The era of working productively with many women in the Republican Party was over. Now the likes of Mary Dent Crisp, a pro-choice, pro-ERA feminist who had served as the Republican National Committee cochair for three years, became the target of relentless attacks by the right. The subject of a stinging rebuke by Reagan at the convention, Crisp was summarily ousted from her position. For the first time, the party platform contained an entire section on abortion, affirming the party's "support of a constitutional amendment to restore protection of the right to life for unborn children." Similarly, a plank of the platform about the judiciary read, "We will work for the appointment of judges at all levels of the judiciary who respect traditional family values and the sanctity of innocent human life."

In November, Reagan and Bush defeated Jimmy Carter and Walter Mondale in a landslide, and when they took office, in January, it was as if the dam had burst. Cultural warfare was at full force. Evangelical broadcaster Jerry Falwell had joined forces with Heritage Foundation cofounder Paul Weyrich to launch the Moral Majority, a powerful political lobby for evangelical Christians that became a "pro-life," "pro-traditional-family," "pro-moral" political-action committee and successfully delivered the Christian evangelical vote for the Republicans. Falwell's Thomas Road Baptist Church in Lynchburg, Virginia, became a megachurch with more than twenty thousand members. Similarly, Falwell's Liberty University became the largest Christian university in the world.

Meanwhile, James Dobson, of *Focus on the Family* fame, a daily radio-show broadcast on more than seven thousand stations worldwide, founded the Family Research Council and began to oversee a family-values empire that included scores of ministries and at least ten magazines that sermonized against gays, the use of condoms, abortion, and the teaching of evolution. Later, Dobson went so far as to attack the Girl Scouts for being "agents of humanism and radical feminism" and to assert that SpongeBob SquarePants, a popular character on children's television, had a secret homosexual agenda. Tim LaHaye and his wife, Beverly, founded Concerned Women of America to counter feminists and to bring "biblical principles into all levels of public policy."

This new wave of reaction was bad news for a wonderful group of moderate Republican women — women like my mother — who believed in Planned Parenthood, were pro-choice, and thought that government had no business getting involved in women's personal decisions. Many Republican women activists stayed loyal to their party. But over time, it began to become clear that they didn't have a future there.

THE ONSET OF THE Reagan years didn't mean that it was time to give up — far from it. With the Windom Fund being administered by Lael, over the next five years I funneled more than $1 million in donations to groups and efforts focused on women's issues and voter registration. They included the Self-Help Center for Women Ex-offenders in the District of Columbia, the Gender Gap Coalition in Chicago, and the El Paso–based *La mujer obrera,* a newsletter for Chicana factory workers — besides, of course, the NWPC and the Women's Legal Defense Fund.

During much of that period, I remained anonymous. Because people constantly tried to pry the identity of the donor from Lael, we created a mythical benefactress — Henrietta C. Windom. The fund was actually named, rather randomly, after a street I'd lived on, but I told Lael, "We'll just say she invented Tampax and wanted to give her money back to women."

One day Lael found a portrait of a young woman who could have been the perfect Henrietta, and we gave her a place of honor by hanging the portrait over the water fountain at the office. All of which was fine — for a while. But soon, a newspaper article men-

tioned the good works of the Windom Fund and, not long afterward, my secretary came into my office and announced, "The Windoms are here."

I was baffled. The Windoms were a figment of my imagination.
But my secretary continued. "There's a couple out front who say
their brother sent them the article, and they want to know if they're
related to some long-lost Henrietta Windom."

At that moment, I realized my anonymity was not going to last
forever. I also realized that if I were to be truly serious about philanthropy and working in nonprofits, I should do it in a businesslike
fashion, so I enrolled in George Washington University to get my
master's degree in business administration. My goal was simply to
get as much bang for my philanthropic buck as possible.

When it came to political activism, my role was somewhat limited during this period. Increasingly, though, the building at 2000 P
Street where the Windom Fund's offices were, near Dupont Circle,
became a center of activism. In those pregentrification days, Dupont Circle, which had declined in the wake of the 1968 race riots, was beginning to enjoy a resurgence of sorts, as urban pioneers
transformed it into a liberal, bohemian, gay-friendly oasis. This, in
a company town — the company being the federal government, of
course — that was largely run by white men in suits who came and
went every four years with each passing administration.

Situated just a few blocks from the giant lobbying firms of K
Street and not much farther from the White House, 2000 P Street
was a seven-story Beaux Arts structure with a used and rare book-
store in it and, upstairs, a host of do-good nonprofits that included
Judy Lichtman's Women's Legal Defense Fund, Ralph Nader's Public Citizen, Voters for Choice, a wonderful liberal gadfly attorney
named Phillip Stern, and, later, the political consulting firm of Marie Bass and Joanne Howes. What they all had in common was that
they were committed to social and economic justice, whether they
focused on race, consumer issues, campaign finance, or women's

rights. Some of these nonprofits were headed by men, but increasingly women were beginning to play leadership roles in nonprofits. That was new.

Lael, of course, was there running the Windom Fund, and I came by regularly, but it wasn't clear to anyone exactly what my role was. I didn't want to be known as someone who had a lot of wealth, but now, it looked like I was merely an adjunct to Lael. It was starting to become uncomfortable and irritating.

Finally, I went to Judy Lichtman's office, which was also on the fourth floor at 2000 P Street. She was at her desk and I sat down on the sofa, and as we talked about fund-raising, she mentioned the support she had been getting from the bank in Boston. By this time, we were such good friends that it was almost a violation of our friendship not to tell her the source of the money.

"I have something to tell you," I said. "I'm the donor behind the Windom Fund."

Judy smiled and almost started to cry. "My God," she said. "That's so wonderful. I'm so glad it's someone I like."

Initially, I had planned to tell just Judy, but my talk with her went so well that I let my role become widely known. Now that I had finally come out of the philanthropic closet, I was able to work openly with other women with inherited wealth and started meeting people in the progressive foundation world.

AS THE MIDTERM PRIMARY season got under way in the spring of 1982, I was one of many women activists who received a marvelous letter from Fredi Wechsler, my predecessor as press secretary for the NWPC, encouraging friends to contribute to the campaign of her old college friend Harriett Woods, who was running as a Democrat for the United States Senate from Missouri.

Harriett had the makings of a great candidate. Tall and stately, with an athlete's grace that dominated on the tennis court, she had served eight years on the Saint Louis City Council. She was then appointed to the state highway commission, and next elected to the Missouri State Senate, where she was a strong supporter of the Equal Rights Amendment. As a relative of Sen. Howard Metzenbaum (D-OH) and a veteran television news producer in Saint Louis, she was savvy and experienced in both politics and the use of media.

Harriett was the only female Democrat running for the Senate that year. As I began to examine her candidacy, I could not help but look at the composition of the 97th Congress. As a Democrat, it was somewhat embarrassing to acknowledge that when it came to representation of women, Republicans were actually ahead of Democrats. The Republicans had Paula Hawkins of Florida and Nancy Kassebaum of Kansas in the Senate, but the Democrats had no one.

I looked back at the history of Democratic women in the Senate. In its entire history since it was founded, in 1828, there had been a grand total of eight women Democrats in the United States Senate. That tiny number was bad enough, but when one drilled down and dug up the details, the story was even worse. The first woman Democrat, Rebecca Felton of Georgia, the widow of a congressman, had been appointed to the Senate in 1922. Her brief tenure — twenty-four hours, mind you — was not the only problem. In an era that was extraordinarily racist, she was off the charts. In one speech, she asserted, "If it takes lynching to protect women's dearest possession from drunken, ravening human beasts, then I say lynch a thousand a week." Not exactly the kind of role model we were seeking.

The seven other Democratic women who had made it to the upper chamber had also been appointed. In November 1931, Hattie Caraway of Arkansas was named to fill the seat left by her husband, Thaddeus, who died in office. She went on to win two terms and set a number of firsts for women. Similarly, in 1936, Rose Long of Louisiana was appointed to serve the remainder of the term of her late husband, Huey Long, who had been assassinated. The following year, Alabama governor Bibb Graves provoked critics by appointing his own wife, Dixie Bibb Graves, to fill the seat left by Hugo Black, who had been appointed to the United States Supreme Court.

And so it went. In 1960, Oregon's Maurine Neuberger ran and won in a special election after her husband died. In 1971, Louisiana's Elaine Edwards and, in 1978, Alabama's Maryon Allen and Minnesota's Muriel Humphrey all made it to the Senate — but by appointment. Not one had been elected in her own right.

That spring of 1982, it seemed the women's movement was a decidedly mixed success. The era of the stay-at-home mom was over. Women were entering the workforce as never before, winning entry to heretofore male-dominated professions such as law and medicine. In pop culture, Aretha Franklin's version of "Respect," originally written and recorded in the sixties by Otis Redding, had taken on new meaning as an anthem of sorts for the women's movement. Movies such as *Alice Doesn't Live Here Anymore, Norma Rae,* and *An Unmarried Woman* ratified the changing consciousness that was sweeping the country.

Yet, in the midst of this whirlwind feminist revolution, the elephant in the living room was that, for all this progress, when it came to the United States as a representative democracy, half the population was still being ignored. In the House of Representatives, the Democrats had elected a handful of rising stars — Bella Abzug, Elizabeth Holtzman, Barbara Jordan, Barbara Mikulski, Shirley Chisholm, Patricia Schroeder, and Geraldine Ferraro — all of whom were beginning to make themselves heard on the national stage. But when it came to the United States Senate, not one single woman in the Democratic Party had ever been elected in her own right.

That, I decided, had to change.

Four

THE VICIOUS CYCLE (AND HOW TO BREAK IT)

WITH HARRIETT WOODS, we were starting from zero, but this was our chance to make history by electing the first Democratic woman to the United States Senate in her own right. As a result, at 2000 P Street, Betsy, Judy, Lael, and I were suddenly fixated on the Senate race in Missouri that season. I immediately sent a check, as did many of my friends and former colleagues from the NWPC.

What happened next with Harriett's campaign was something that had happened countless times within both major political parties in the United States, which were run by an old boys' network, as it was known in the women's movement. Because we were participating in the political process directly for the first time, we were able to experience the power of the entrenched male establishment in a way that resonated deeply and had a lasting impact. We saw firsthand that the obstacles that arose for women who ran for higher office were a result of a system created by the men in charge in which it was virtually impossible for women to win.

Harriett was a terrific case in point. Even though she had everything it took to be a strong candidate, she was considered a long shot. The reason was simple: gender. The general population was just getting accustomed to the fact that women could be lawyers, so when it came to the United States Senate, they had few reference points to judge a woman's worthiness. After all, what did a woman senator look like? Was any given woman candidate smart enough? Tough enough? Could she juggle her duties as a senator with her responsibilities as a wife and mother? Who would buy the groceries? Who would take care of the kids?

Nevertheless, in August, with no support from the party regulars in Missouri, Harriett took on a field of ten for the Democratic Party nomination for the Missouri senatorial seat. Her upset victory earned her national attention.

Looking toward the general election in November, however, she was still a huge underdog. For one thing, her opponent, John Danforth, was an incumbent, which was an enormous advantage. In addition, because he was a moderate Republican who was against the death penalty, Danforth was in a strong position to win votes from independents and so-called Reagan Democrats. As the heir to the Ralston Purina fortune, he was a Missouri blue blood. Danforth was a staunch Reagan supporter, as well as an ordained Episcopal priest who had been a fixture in Missouri politics since he was elected attorney general of the state at the age of thirty-two. In late September, five weeks before the election, one poll showed Danforth with a seemingly insurmountable 17-point lead.

But Harriett refused to give up. Because she was the only Democratic woman running for the Senate, she became the standard-bearer for feminists that year. She won backing from the National Organization for Women, the National Abortion Rights Action League (NARAL), the Women's Campaign Fund, Friends of Family Planning, the American Nurses Association, the International Ladies' Garment Workers' Union, and the NWPC.

I was in grad school at the time, but many of my friends at 2000 P Street began to meet informally, and they went to work. "There was Lael Stegall from the Windom Fund, Joanne Howes, and Marie Bass," said Betsy Crone, then a fund-raising consultant. "We were all political activists working together. We just began to call friends and raise money. We didn't touch the money — we just sent it on directly to the Woods campaign."

Harriett campaigned relentlessly. Thanks to her experience in TV news, she was adept at getting her message across in short sound bites. Using a slogan — "Give 'em hell, Harriett" — that played off the famous catchphrase of fellow Missourian Harry Truman, she mounted countless attacks against Reagan's policies, stood up staunchly for abortion rights, and drew impressive support from women. She put together enormously effective TV commercials portraying Danforth as a patrician hypocrite who spoke kindly about the working man but voted for big corporate interests. Even with the Saint Louis Cardinals in the World Series — and winning — she repeatedly managed to steal front-page headlines. She picked up an endorsement from the *Saint Louis Post-Dispatch*. By October 15, the *Saint Louis Globe-Democrat*'s poll reported that Woods and Danforth were tied, each with 47 percent of the vote.

Danforth was stunned. "There I was, breezing around the state, Mr. Nice Guy," he said. "Then all of a sudden . . . it was like running into a brick wall." Now that he finally felt the heat, Danforth resorted to calling Harriett "a demagogue."

As the campaign entered the home stretch, Betsy Crone, Joanne Howes, Lael Stegall, and I were on tenterhooks. Harriett might win — and we were calling friends and sending on as much money to her campaign as possible. But it wasn't enough. According to the Associated Press, Danforth raised more than three times as much as Woods, and on October 15, the same day she pulled into a tie with Danforth, she ran out of money.

Even though some polls actually showed Harriett in the lead, the Democratic Party offered her little or no support. They couldn't believe a woman could beat John Danforth — even though she was the nominee and was tied in the polls. As a result, even though the TV ads were clearly working, during the last two weeks of the campaign she had to yank all of them, because she was broke.

Finally, on Election Day in November, Harriett lost by a mere 26,200 votes out of 1.5 million. "I can't prove it, of course," said her pollster, Irwin "Tubby" Harrison, "but I really do believe if we had been able to stay on the air continually, we would have passed [Danforth] in mid-October and won the race."

THE LESSON OF HARRIETT'S loss was painful but immediately apparent to all of us. Because the men who ran the Democratic Party were convinced that women could not win, they withheld their money. In effect, they had created a system that produced self-fulfilling prophecies time after time, a system in which even highly qualified women candidates were bound to lose. "Watching Harriett lose taught us a big lesson," said Judy Lichtman. "We realized that business as usual was not going to get us anywhere, and we had to do things differently."

Rather than stew over this sorry state of affairs, we did something about it. In the winter of 1983, Lael suggested I host a breakfast for a small group of women who had experience in politics: Betsy Crone; Judy Lichtman; Joanne Howes; Marie Bass; Jane McMichael, who was executive director of the NWPC when I worked there; Joan McLean, who was political director of the caucus; and Ranny Cooper, executive director of the Women's Campaign Fund, a nonpartisan organization for providing resources for women candidates.

We met in a private room upstairs in the Tabard Inn, a quaint tavern near Dupont Circle that is an iconic Washington haunt reminiscent of an English manor, with cozy fireside tables and a maze of alcoves. We were fairly young back then — most of us were under forty — but all of us had had years of political experience in a variety of liberal nonprofits and Democratic Party organizations, ranging from George McGovern's 1972 presidential campaign to Teddy Kennedy's in 1980, Common Cause, the NWPC, senatorial staff jobs, and political consulting. These were the women who shared my values and interests and who later became known, along with Millie Jeffrey and others, as the Founding Mothers of EMILY's List.

After Harriett's loss, our goal was clear: we wanted to elect a woman to the United States Senate. That, we realized, would be a monumental task. By and large, except for a handful of congresswomen, when it came to big-time politics, women were nothing more than envelope stuffers. Washington was essentially a men's club. On the rare occasion when the Democratic Party asked a woman to be a candidate, it usually was a fool's errand or, more likely, a kamikaze mission — to run as a hopeless long shot, for example, against a popular incumbent Republican in an overwhelmingly Republican district. There were hundreds of advocacy groups in Washington, and many of them did fine work. But I had something different in mind. Creating progressive policies and promoting them can be incredibly valuable. But those policies will never be implemented unless enough politicians are elected who support them. To that end, I wanted to hone in on what would be the single most effective strategy to put female, pro-choice, Democratic women politicians in office. Now, with Harriett Woods, we had first seen exactly what we were up against: an entrenched male political establishment that had created and perpetuated a system full of obstacles that made it almost impossible for women to get elected.

After we sat down and placed our meal orders, I posed a series of questions: What do we need to do to convince the establishment that women can win? How do we boost the credibility of women candidates so the establishment will support them? If the Democratic Party gave such short shrift to someone as well qualified, as highly electable, as Harriett Woods, how do we change that? We needed a simple, clearly defined strategy that would overcome the systemic obstacles we faced.

We went around the table, and without exception we all agreed that it came down to one thing: money. Specifically, we'd learned enough to know that *early* money had enormous strategic value that could be leveraged into real credibility that would get additional funds later on. If our candidates had more seed money, they could begin building a campaign organization, field their own polls, and create a strategy for victory — all of which would convince the traditional funders to support their efforts.

A number of our allies — NOW, the Women's Campaign Fund,

NARAL, Planned Parenthood, and the NWPC had political-action committees (PACs). But none of these PACs were well funded or could give much support. Even if they had had resources, PACs were restricted from giving more than $5,000 to any single candidate per election — not nearly enough to be decisive in a senatorial race. To make matters worse, big-time women political donors were all but nonexistent. Even though by now women professionals were entering the workforce, in many important ways it was still a man's world. Women were not yet used to handling the family checkbook, much less to making political donations. Men at the big law firms had no problem writing checks for $1,000, which was then the legal limit per donor per candidate, but to the extent that women made donations, they wrote checks for $10 or $20. In her House campaigns, Barbara Mikulski had to resort to dozens of nickel-and-dime efforts to get women to support her. There were Bake Sales for Barb, Barbecues for Barb, and Baseball for Barb. But all she got were small checks.

To overcome these obstacles, we decided on a very simple plan. None of us really wanted to start a new, full-fledged political organization, so we decided simply to write letters to people about the Democratic women who were running for the Senate who we thought had a good chance of winning. We would suggest that the letter recipient write checks to the candidates we described, send them directly to the campaigns, and pass our letter on to others. We had in mind a political chain letter that we hoped would result in an avalanche of support for each candidate.

One more thing. Even though we were not a formal political organization, our group wanted a name that would be catchy and reflect our commitment to raise early money for the candidates. Nobody had any compelling suggestions, so we tabled that for later discussion.

But one day, I walked over to Lael's office so we could review the agenda for our meeting later that day. While Lael finished a phone call, I began playing with the phrase "early money." That was the strategic imperative we all agreed on. I began thinking in terms of acronyms. Early money. *EM.* "Emily," I thought. One of my close friends in grade school was named Emily.

"Is." That took care of the *I.* Now I needed an *LY.*

And then it just happened: "Early Money Is Like Yeast." After all, what does yeast do? It makes dough rise.

EMILY's List. The idea of a list had a certain cachet. It suggested exclusivity, something one wanted to be a part of. Also, given that we were situated in a world that was drowning in federations, coalitions, associations, alliances, and such, it created an identity that contrasted nicely with the slew of bland and predictable acronyms that flooded Washington. And it spelled out our strategy for all to see.

When Lael finally got off the phone, I told her about my latest idea. "How about EMILY's List: Early Money Is Like Yeast? It makes the dough rise."

"I love it," she said. "It's absolutely perfect."

We moved into the conference room where the founders were assembled. For an hour, we discussed how to send out our information, where to get names, and who would call the various campaigns of women candidates to find out what was happening. Everyone had taken her assignment, and the meeting was coming to a close — but I hadn't mentioned my idea for the name.

Then, just as we were about to disband, Lael spoke up. "Ellen," she said. "Tell them about the name."

By this time, I had chickened out. What kind of name was EMILY's List? I'd never heard of an organization called a list. And yeast? *Oh, I don't know,* I thought.

But Lael egged me on, so I finally relented. "EMILY's List: Early Money Is Like Yeast. It makes the dough rise."

And they loved it. EMILY's List had a name.

Five

THE FERRARO FACTOR

AS THE 1984 ELECTION SEASON got under way, what was most exciting was the possibility that a woman might win a place on the national ticket for the first time in history. In large part, that was thanks to what was known as Team A, a savvy group of activist women including Millie Jeffrey, Joanne Howes, Ranny Cooper, and Joan McLean.

But a woman vice president? At the time, even some of the more successful women in Congress, such as New York's Geraldine Ferraro, thought the idea was far-fetched. Still, women had been winning a foothold in the legal profession for more than a decade. In 1981, President Reagan had appointed Sandra Day O'Connor as the first woman to sit on the United States Supreme Court. By this time, twenty-four women were serving in Congress — including two in the Senate — and some had already won national recognition.

With women's groups all over the country calling for a female veep, Team A's role was to implement a behind-the-scenes strategy that would turn the public pressure into reality by focusing on one specific candidate. So, Team A went into action and discreetly started to vet some of the most prominent women in Congress.

Actually, to call this effort discreet is to make a bit of an understatement. "We didn't want people to think we were making any of this happen," said Joanne Howes. "It was guerilla warfare. The entire women's movement was talking about a woman vice president, but we weren't interested in a merely symbolic campaign. We wanted to be sure that a woman would be nominated. That meant seriously vetting potential candidates."

Early on, Team A knocked out Mary Rose Oakar of Ohio and Lindy Boggs of Louisiana, because they were anti-choice. Pat Schroeder of Colorado didn't make the cut because the paucity of her state's eight electoral votes meant she did not bring much value to the ticket. Rep. Bella Abzug talked to Barbara Mikulski of Baltimore, former representative Barbara Jordan of Texas, and others. San Francisco mayor Dianne Feinstein was under consideration, but she was divorced, Jewish, and from San Francisco, three factors that did not play well in the heartland.

To my surprise, the higher-ups in the Democratic Party were more receptive to the idea of a woman as a vice presidential candidate than we had dared hope. Former vice president Walter Mondale, the odds-on favorite to win the Democratic nomination, thought that a woman vice presidential candidate might be just the ticket to inject some excitement into the Democrats' campaign. President Reagan had a commanding lead in the polls, so there was a growing sense that the Democrats had to do something dramatic. On September 24, Mondale told the Americans for Democratic Action that he wanted to "bring women into positions of power like they've never been before."

By the fall of 1983, Team A had started to look more closely at Rep. Geraldine Ferraro (D-NY), a three-term congresswoman. A member of the first wave of women who had entered the law, Ferraro, an Italian American from the borough of Queens, had won a reputation as a tough but fair prosecutor while serving as assistant district attorney in the seventies. Ferraro was not exactly a household name, but Team A had written up notes describing each potential candidate, and its assessment of her indicated she was among the least problematic contenders: "three term congresswoman; married and mother of three; friend of labor, the elderly, and women; respected by House leadership; East Coast blue collar; conservative ethnic constituency."

In November, Joan McLean, the former political director of the caucus, approached Ferraro's administrative assistant, Eleanor Lewis, and told her a small group of women would like to meet with Gerry about running for vice president.

Lewis was stunned. "Vice president of what?" she asked. "Of the United States?"

Not long afterward, Ferraro and Lewis met over a dinner of Chinese food in Washington with Team A members Nanette Falkenberg, executive director of NARAL; Millie Jeffrey; Joan McLean, a staffer on the House Committee on Banking; and Joanne Howes.

The team members came right to the point. Team A wanted to promote Ferraro as a candidate for the vice presidency. Ferraro was both stunned and flattered by the idea. "Our intention was only to get Gerry's agreement that if she were nominated, she would serve," said Joanne. "She had to be willing to have this thing roll out, and see where it went."

In response, Ferraro said she was willing to be a candidate, but she felt that getting the nomination was a long shot for any woman. Then, as the meal came to an end, she read the prophecy in her fortune cookie: "You will win big in '84," it read.

"I can't believe what we just talked about," she told Eleanor Lewis on the way home. "Am I the only woman in Washington who doesn't think a woman can get the vice presidential nomination?"

BUT TEAM A WAS NOT ALONE in thinking that a woman might be right for the number-two spot. By late spring, after Walter Mondale had triumphed in the primaries, the option of a woman veep became increasingly likely thanks to a secret hundred-page strategy plan drawn up by Mondale field director Mike Ford. According to Ford's document, Mondale's chances against the Reagan-Bush team were disturbingly bleak, so it was essential to consider "dramatic and perhaps high-risk strategies."

Meanwhile, Mondale had settled on "bold choice" as a catch-phrase for choosing a minority or a woman candidate for the number-two spot. Given that there was little suspense about who would win the Democratic presidential nomination, the prospect of a female vice president fueled the media throughout the primary season. On June 4, San Francisco mayor Dianne Feinstein and Ferraro graced the cover of *Time* magazine under the cover line AND FOR VICE PRESIDENT . . . WHY NOT A WOMAN?

In the *Washington Post,* Mikulski joked about it. "When Pat Schroeder, Geraldine Ferraro and I get together," she said, "the only thing we don't agree on is who should be vice-president."

By this time, however, Ferraro had caught the eye of Tip O'Neill, the powerful and candid Speaker of the House. "She's one of us," he confided to colleagues, suggesting that Ferraro was a real pol who knew how to play the inside game. Coming from O'Neill, that was high praise indeed.

O'Neill, a crusty, old-style Boston Irish politician, passed the word to Mondale. "She's a star," he said. "If you're looking for a VP, she's the one." Soon, O'Neill went public. "Sure, I have a candidate," he told the *Boston Globe.* "Her name is Geraldine Ferraro, she's from New York, she's a Catholic and she's very smart." It was, as Ferraro later put it, "the *Good Housekeeping* seal of approval."

On June 30, less than two weeks before the 1984 Democratic National Convention, Walter Mondale arrived in Miami Beach to address the National Organization for Women. By this time, the ERA was dead, because time had run out in the ratification process. Instead, NOW had made getting a woman on the ticket one of its top priorities. NOW members sported buttons proclaiming MON-DALE-FERRARO, MONDALE-FEINSTEIN, and MONDALE-MIKULSKI. Workshop sessions titled "The Gender Gap" and "Why and How Women Will Elect the Next President" generally began with the premise that Mondale needed women to win — and women wanted a spot on the ticket in exchange for their support.

Meanwhile, various other potential vice presidential nominees fell by the wayside. At age sixty-six, Los Angeles mayor Tom Bradley was judged to be a weak campaigner. San Antonio mayor Henry Cisneros, a thirty-seven-year-old Hispanic, was too young.

As for Ferraro, what most appealed to Mondale was that she easily embraced both the old and new traditions of the Democratic Party. She was enough of a pol to smile gracefully when Tip O'Neill called her "hon," but she was equally comfortable with the new wave of feminists. And she had a disarming natural warmth that enabled her to be decisive and tough-minded without alienating people.

By Wednesday, July 11, Mondale had made his decision. Ferraro was at a hotel in San Francisco when she got his call. About five minutes later, when their talk was over, Ferraro opened the door and talked to a staffer. "How does it feel to be a part of history?" she asked.

THE NEXT DAY, JULY 12, I turned on the TV to watch Mondale's press conference. With Ferraro at his side, Mondale said, "Today, I'm delighted to announce that I will ask the Democratic National Convention to nominate Geraldine Ferraro of New York to run with me for the White House."

"This is an exciting choice," Mondale continued. "I want to build a future . . ." Then, he paused. Mondale was not a man who expressed feelings easily; nor was he adept at eliciting emotional response. But the crowd erupted at the sight of him and Ferraro together. Even jaded reporters were filled with emotion. Mondale was silent for a moment, grinning from ear to ear with pride, as if fully realizing, for the first time, the momentousness of the historic occasion.

"Let me say that again," he said. "This is an exciting choice . . ."

I watched with a mixture of excitement, pride, and anxiety. I thought, "Oh, Gerry, don't cry! Boy, I bet I would cry. Oh, this is so fantastic."

There were other politicians whom I had eagerly voted for and admired, but this was different. I connected with Geraldine Ferraro.

For the first time, I could identify personally with a candidate for major political office. Geraldine Ferraro was a woman, just like me. I was so proud of her.

Four days later, when the Democratic National Convention assembled in San Francisco, Ferraro was nominated by acclamation in a voice vote. "We will place no limits on achievement," she said, in accepting the nomination. "If we can do this," she added later, "we can do anything. We must not go backwards."

"It was one of the most electric moments in American politics," said the *Wall Street Journal*'s Al Hunt. The public reaction was incredible. The following week, Gerry was on the cover of both *Time* and *Newsweek*.

For millions of women, it was a deeply moving experience. "It was like history washed over them," said Marie Wilson, president of the Ms. Foundation for Women. "To not see your personage reflected in the public world as a woman was a pain. And when they saw her on stage or in person, it was like a healing."

"Even my Republican friends were thrilled that a woman was the vice presidential candidate," said Hillary Clinton. "There was this sense of it being a watershed, an historical turning point."

NOR DID GERRY let us down. A relative newcomer on the national stage, she came off as attractive and quick-witted, tough without being abrasive. A week after the convention, the Mondale-Ferraro ticket, which had been 18 points down, pulled within striking distance of the Reagan-Bush team. In *Newsweek,* a Gallup poll even showed the Democrats pulling into a 2-point lead. Now that I had finished my studies for my MBA, I eagerly volunteered for the Mondale-Ferraro campaign.

As the campaign got under way, "the Ferraro Factor," which had initially been deemed a big risk, instantly became a big plus. Throughout the campaign, gender intruded into matters both trivial (should Mondale kiss Ferraro on the cheek?) and material (should husband John Zaccaro release his tax returns?). But in general, the enthusiasm was extraordinary. All over the country, hundreds of thousands of women who had never been political suddenly turned out to see Gerry. Young mothers — and fathers — attended rallies with their infant daughters, whom they hoisted aloft to view Gerry as if to imprint on their progeny the notion that young women could grow up to be whatever they wanted to be.

Ferraro handled often-awkward gender issues with style and grace. As the *New York Times* pointed out, for the first time voters heard a vice presidential candidate address the issue of abortion with the phrase "if I were pregnant," and foreign policy with the phrase "as the mother of a draft-age son." Her acolytes responded by wearing FERRARO-MONDALE buttons, proudly inverting the hierarchy of the Democratic ticket as a commentary on which candidate was truly the most important. Suddenly, it was apparent that Gerry could bring women out who had never voted before and could become a powerful electoral force.

But the men managing the campaign weren't interested in including women's voices in the decision making. Newly anointed as a heroic, history-making candidate, Ferraro herself found that her role as a "partner" in the campaign was nominal at best. "We were in North Oaks [Mondale's Minnesota home], meeting with Mondale and his staff, and they made a presentation of what the campaign would look like over the next four months," recalled Eleanor Lewis in *Geraldine Ferraro: Paving the Way,* a documentary. "And it was in ink."

"Why is this in ink?" Gerry said. "Why isn't it in pencil? Why am I not being asked before it's put in?"

She got up and stormed out. She insisted that she be a partner in the decision making. But that was not to be. "In theory, they redid the plans," said Joanne Howes. "But in reality they didn't. It stayed in ink."

Similarly, Team A, having orchestrated Gerry's nomination, found itself marginalized. "The day she got nominated, we were shut out completely," Joanne continued. "This was Mondale's campaign. He had picked Gerry, and he was going to pick the people who were going to work for her, and that's the way it was. They didn't really understand why they nominated her. They didn't 'get' the power of the women's vote."

The Republicans, however, *did* realize a historic moment was at hand. "The other side saw the power of the Geraldine Ferraro candidacy, and they had to try to diminish it as soon as possible," said Nancy Pelosi, then chair of the California Democratic Party.

Suddenly, the adulatory coverage of Ferraro gave way to an orchestrated campaign of damaging leaks about her husband's finances, critical stories about her by anti-abortion forces, and criticism from the Catholic hierarchy. "There was a covert operation, no question about it," said Ed Rollins, the Reagan-Bush national campaign manager, in *Geraldine Ferraro: Paving the Way.*

When Gerry said she didn't believe Reagan was a good Christian, because his social policies were "so terribly unfair," Nancy Reagan had had enough. She tasked Reagan operatives with planting aggressively negative stories about Ferraro in the press. As a result, the postconvention bounce following Ferraro's nomination vanished, and by early October, the Republican ticket had a lead of more than 15 points in some polls.

All of which made the October 11 Ferraro-Bush vice presidential debate between Ferraro and George H. W. Bush both an epic gender battle and a David-Goliath clash pitting Bush's imposing résumé — congressman, UN ambassador, head of the Republican National Committee, envoy to China, head of the CIA, and vice president of the United States — against that of a three-term congresswoman who was not even nationally known.

As I watched the debate, I thought Gerry was terrific. Smart, informed, personable, and comfortable, she more than held her own. Declaring that any Soviet nuclear threat "would be met with swift, concise, and certain retaliation," Gerry proved that a woman could be tough on foreign policy. Likewise, her lightning-quick "don't-you-dare-patronize-me" retort to Bush's condescending reprimand

("Let me help you with the difference, Miss Ferraro, between Iran and the embassy in Lebanon.") showed she could stand toe-to-toe with the best of them on the world stage. "Let me, first of all, say that I almost resent, Vice President Bush, your patronizing attitude that you have to teach me about foreign policy," she shot back.

Critics who had doubted whether she — or any woman — had the temperament, experience, and fortitude to hold such a high office suddenly had nothing left to say. In the end, Bush had to fall back on a bit of faux machismo the following day — "We tried to kick a little ass last night" — and wife Barbara was reduced to prissy, sophomoric name-calling — "I can't say it — but it rhymes with rich." All of which told me that Gerry had delivered something that was absolutely critical for women politicians: credibility.

FOR ALL THAT, by any measure, Election Day, Tuesday, November 6, 1984, was a catastrophe for the Democratic Party. In one of the greatest landslides in history, the Reagan-Bush ticket won the popular vote by seventeen million votes and took forty-nine out of fifty states,* winning 525 electoral votes out of 538, the highest total ever received by a presidential candidate.

Even if Ferraro wasn't the problem — and she wasn't — she alone wasn't the solution either. "Ferraro has helped," the NWPC's Monica McFadden told columnist Maureen Dowd of the *New York Times*. "There has been more emphasis and higher visibility for women of both parties. But will Gerry's apron strings . . . carry people into office? No."

So, even though seventeen Democratic women challenged incumbents in Congress in 1984, not a single one emerged victorious. As for the EMILY's List chain letter we sent out, the results were no better. All four of our senatorial candidates lost, and only one, Joan Growe in Minnesota, topped the 40 percent mark. "They were all

---

* The only state that the Democrats won was Minnesota — Mondale's home state — and that they carried by just 3,761 votes.

running in races where Democrats didn't stand a chance in hell of getting elected," said Joanne Howes. "So, yes, they let a woman get the nomination—but only when it was to run against Republican incumbents who couldn't be beat."

So much for the "magical breakthrough" many women had longed for. As far as EMILY's List was concerned, the 1984 campaign was essentially a dry run. Betsy Crone had pulled together as many mailing lists as she could. Joanne Howes and Marie Bass had served as our political team assessing a fairly weak slate of pro-choice Democratic women running for the Senate in the 1984 elections. I had written a letter that talked about our mission to elect the first Democratic woman senator and our early-money strategy, and asked people to join as "captains" who would each recruit four other donors who would contribute from $250 to $1000 to each candidate. "Since we are confining ourselves to women Democratic senate candidates who are progressive feminists, we expect to inform you about two to five candidates in any two year election period," the letter read. "I want to emphasize that this is a network for early money. Unlike the 'old boys' network,' we are willing to take a chance on women candidates." We had urged each reader to send checks directly to the campaigns, making clear they were part of EMILY's List.

People had responded with excitement. But because we were raising money only for the candidates—not EMILY's List—the contributions were not coming to us, and because others were passing the letters along, in good chain-letter tradition, we had no way of tracking how successful we were, or even who actually was participating. A vital part of success in campaign fund-raising is building lists so you can go back to donors. But we didn't know who had contributed, and so we weren't in a position to solicit the same contributors again for the next election cycle.

ALL IN ALL, ONE might have thought 1984 was enough to have killed EMILY's List in its cradle. But I saw things very differently

— as did Joanne, Millie, Betsy, Lael, Judy, and my other friends. Yes, 1984 was a staggering defeat for the Democrats. But for us, it was also a staggering opportunity. That's because, after taking the historic step of nominating Gerry, the Democratic Party leaders had not seen what was staring them in the face. All across the country, at one stop after another on the campaign trail, hundreds of thousands of women showed up for Gerry. Tens of millions of women had voted — half of the entire electorate. How could the party elders be so blind?

And women had the potential to do more than just vote; hundreds of thousands of them were potential donors, a completely untapped source of funds that the Democratic Party barely bothered to explore. "I kept thinking, why isn't the campaign raising money from these people?" said Joanne Howes. This huge opportunity to raise money from women and to learn better how to target women voters was wasted — to the detriment of the party in 1984 and beyond.

And, finally, there was a far smaller group consisting of women like me, Joanne, Marie Bass, Joan McLean, and Betsy Crone who had been instrumental in discreetly getting Gerry Ferraro nominated, and other women such as Nikki Heidepriem, Carol Tucker Foreman, and Mimi Mager, whom I'd met working for the Mondale-Ferraro campaign, who shared our values and goals and joined in soon afterward. These women were not merely feminists. Nor were they ideologues. They were savvy political professionals who were young and dedicated and had already mastered a wide range of political skills, who had expertise in public relations, campaign fund-raising, direct mail, and political strategy, who understood the legislative process, who had staff experience with the Democratic senatorial leadership, and who could navigate the corridors of power in Washington. And yet somehow they had been shunted aside by the old boys' network of the Democratic Party, their advice ignored or excluded completely from deliberations in the male-dominated world of political campaigns.

The dissonance was extraordinary. Even after the party leadership — virtually all male, of course — finally took that historic step of nominating Ferraro, they went back to business as usual. "There had been this incredible sense of euphoria when Gerry was nominated," said Joanne. "A barrier had been broken. But when they didn't follow through, it was like a punch in the stomach. They couldn't see the significance."

So, while the party elders sat around licking their wounds after Election Day, my friends and I were focused on the chasm between the millions of energized women looking to make a difference and the way the old boys' network ignored them, the gulf between the potential and the reality — and how to bridge it. Ultimately, the real lesson of 1984 was that we hadn't gone far enough. We needed to take what we had started with EMILY's List and build it into a powerful network to raise early money.

In the end, rather than rant and rave at the leadership of the Democratic Party, I almost wanted to thank them for allowing us to take advantage of their failure to harness the power these women represented. After all, they had provided us with a spectacular opportunity. And we weren't going to let it pass.

Six

THE FOUNDING MOTHERS

AND SO IT BEGAN. Almost immediately after the debacle of 1984, I started talking to Lael Stegall and some other friends about pouring all my professional energies into EMILY's List. My MBA contributed to what I knew about marketing. At Common Cause, I had helped build an organization from scratch. I also had the beginnings of a strong support network: Lael was doing a terrific job with the Windom Fund; Betsy Crone's days at the NWPC had honed her skills at direct-mail fund-raising — how to assemble mailing lists, write solicitation letters, and handle production; Judy Lichtman, my buddy across the hall at 2000 P Street, was my constant adviser, and I had joined the board of her Women's Legal Defense Fund; Joanne Howes, whom I knew from my days at the caucus, as well as her consulting partner, Marie Bass, were deeply experienced in the political process. There were others as well, including Teddy Kennedy aide Ranny Cooper; Ann Kolker, with whom I shared an office, and Mary Ann Stein, whom I met through my Windom Fund work. As the group coalesced, I realized we had put together a powerful assemblage of smart, committed, highly professional, highly motivated women.

I brought in Gail Harmon, an election-law specialist who had worked with the Windom Fund, to transform EMILY's List into a political-action committee, albeit a completely different animal than most PACs. Most PACs, after all, were organizations funded by businesses, trade groups, labor organizations, or individuals with ideological or legislative agendas — tobacco, oil, health care, labor, and the like — and the money they raised generally gave them access to legislators to promote their aims. Our purpose was very different. We didn't want access to members of Congress to promote a legislative program. Our goal, pure and simple, was to elect pro-choice Democratic women.

But going this route meant we had to circumvent a major potential problem: PACs could legally contribute only $5,000 to a given candidate, and even in those days, before the advent of the super PAC and billionaires like the Koch brothers, $5,000 was a drop in the bucket. So, if we were going to make EMILY's List work, we had to come up with a strategy that was dramatically different from that of other PACs. We had to revolutionize political fund-raising.

Doing that meant creating a unique kind of PAC: We would bundle money from thousands of women, in effect launching a donor network with a shared goal of raising early money for Democratic women candidates. Instead of focusing on raising money for EMILY's List itself, we would build a donor network of women and men who were committed to giving to candidates we recommended. As a PAC, EMILY's List could contribute only $5,000 per election. By operating as a donor network instead, we could raise virtually unlimited amounts of money.

In a way, it helped that we were so different from the kind of PACs one finds among the K Street lobbyists, and it made sense to use that to our advantage. Our chances of electing lots of pro-choice women would be infinitely greater if we could harness the immense untapped collective power of hundreds of thousands of people, overwhelmingly women, who had worked their hearts out for Harriett Woods and who had been so excited by Gerry Ferraro's nomination. They were well-educated, affluent professionals in law, medicine, and other professions who were deeply committed politically but had been overlooked and marginalized by old boys' networks all over the country. We were creating a vehicle through which these women would have a political voice.

In effect, it was a more formalized version of the chain letter — through which we reached out to people we knew to get them to write checks to the candidates, and to get their friends to do likewise. Our major goal was to be fund-raisers for the campaigns, but in order to do that effectively, we needed to have a formalized organization so we could raise money in a far more professional way than we had done in 1984.

And we didn't even know whether a donor network was legal.

Betsy Crone noted that the Council for a Livable World, a nonprofit for nuclear nonproliferation, had done something along these lines, so Gail Harmon checked and assured us that as long as the individual donors controlled who the recipient was, we could function legitimately as a donor network and bundle as many checks as possible. When this hurdle was passed, we knew the sky was the limit for what we could raise.

Finally, in December 1984, Gail came over to the Windom Fund's office to meet with Lael and me to sign the legal papers. We went through the ritual of signing one document after another, officially registering EMILY's List as a political-action committee.

Then, I got out my checkbook, and tore out two checks. The first I dated late December, and made it out to EMILY's List for $5,000, the maximum an individual could contribute to a political-action committee at that time. I then wrote a similar check for January as well, and handed it to Lael.

EMILY's List was under way.

WE DIDN'T WASTE any time. The next month, well in advance of the '86 electoral cycle, the Founding Mothers began meeting regularly, usually in the fourth-floor conference room that the Windom Fund shared with the other nonprofits at 2000 P Street.

More than anything, I wanted to make EMILY's List a success, but I was also somewhat tentative, because leadership was still new territory for me. I had known most of the founders for some time, but many of them had been in book groups, played tennis, and socialized together, so their ties to one another were deep. I knew I had to prove myself to win acceptance as their leader. The year before, when we had met at the Tabard Inn, I had been the host, coordinating the discussion and asking people what they thought. Now, we needed to do more than talk. If we didn't start now, if we didn't act, everything would just blow away like smoke. People would come to the meetings one or two times more and then lose interest.

At one of the very first meetings in January, Judy Lichtman raised an important practical question: Why would anyone give to our organization if they didn't know who we were? We weren't exactly household names. Joanne Howes suggested that we create an advisory board of prominent women and put their names on our letterhead.

I swallowed hard. It was time to put aside my hesitancy. It was time to lead.

"Great idea," I said. "But I don't think we should spend a lot of time on this, so let's set a deadline for three weeks."

Then, I asked Joanne to call Ann Richards, a rapidly rising star in Texas politics who had just been elected state treasurer. Likewise, I suggested that Judy Lichtman reach out to Donna Shalala, then president of Hunter College. These were the names that would go on the letterhead of the first solicitation letter we sent out. One by one, I made assignments.

All around the table, heads nodded. There was a group consensus that I was our leader. That was the moment I effectively became president of EMILY's List. And with an action plan being implemented, we were now on our way to transforming fantasy into reality.

Early on, we decided on our goals. We wanted to build a network of at least one thousand donors who would contribute $100 each to join EMILY's List and then would agree to donate $100 each to two or three candidates we would recommend. (That $100 figure may seem low today, but at the time the conventional wisdom was that it was too high.) Altogether, we hoped to raise $100,000 each for our candidates. Finally, the bottom line was that we wanted to send the first Democratic woman elected in her own right to the United States Senate.

We were in uncharted territory when it came to fund-raising. Often, in direct-mail fund-raising, you lose money on the first mailing but you survive financially by going back to the original donors and resoliciting them. All direct mail is based on that premise.

But we were doing something very different. We were asking donors to join us. And then we were asking them to trust us enough to give money not to us but to the candidates we recommended. That left two big unanswered questions: first, could we raise enough to keep the doors open, given that we were not resoliciting for ourselves, and second, could we build enough trust so that our donors would actually make out checks to the candidates we endorsed? Everything else paled in comparison to getting this concept right. Nobody knew whether it would work.

All of which meant we had a significant marketing task ahead of us. The unique name — EMILY's List — helped make it clear that we had a product that was unlike anything else. The name showed that we didn't take ourselves too seriously but that we were serious about electing women. The idea of building trust with members was absolutely critical, so it was essential that members could see exactly where their money was going, that we be completely transparent. In the end, we felt we could implement the fundamental rules of modern marketing: you have the need, we have the solution, so choose us as your product.

One of our first tasks was to define our market. We didn't have the money to do the kind of market research usually done in the corporate world, but figuring out our market segment wasn't terribly difficult. We primarily sought women between the ages of thirty-five and sixty-five who were well educated, political, and used to supporting causes they cared about, such as the Equal Rights Amendment. They had given financial contributions to political causes and/or worked for them. They identified themselves as pro-choice feminists who were progressive on a whole host of issues and were frustrated that our representative democracy excluded elected officials who looked like them.

In other words, we, the founding mothers, *were* the market segment.

Betsy Crone knew where to find our potential members. They were donors to women's organizations, the Democratic Party, Planned Parenthood, and Democratic candidates, especially women candidates. So, she began to collect or rent direct-mail lists from those organizations. "In those days, it wasn't as if you went to school and got a degree in fund-raising," Betsy recalled. "There was no one teaching us these techniques. We just reached out to women all across the country because these were issues we cared about."

Meanwhile, I set up meetings with the heads of national women's political and advocacy organizations to talk about our plans. Because it was unusual for organizations to share their donor lists of $100 contributors, I thought it was important that I personally ask for help from these leaders. They were generous and supportive. I was particularly grateful to Molly Yard, vice president of NOW, who responded with sisterly generosity whenever I called or visited.

It would be an understatement to say that ours was a lean operation. In addition to my contributions, we started out with $60,000 in seed money we raised from a small group of donors I knew and progressive groups including the National Education Association; the American Association of State, County, and Municipal Employees; the Communications Workers of America; and Sen. Edward Kennedy's Fund for a Democratic Majority.

Initially, the total number of salaried, full-time staff at 2000 P Street was exactly zero. But that didn't mean we didn't get things done. As Betsy put together mailing lists of donors to women's organizations, Marie Bass and Joanne Howes organized parties at which we told people about EMILY's List and asked them to join.

They also became our key political advisers. Cat Scheibner, an early staffer, worked part-time and made sure our computers and our office systems were running smoothly. Janet Corrigan, who had been the Windom Fund receptionist, kept our finances in order. We hired Kathleen Currie as our press secretary and writer.

Meanwhile, I served as EMILY's List's unpaid president and general manager. I spent much of my time working with Betsy as she supervised direct mail and administrative functions while I focused on raising money, marketing, and our message — how to define our market segment, how best to introduce the idea of EMILY's List, what our written materials should say, and what they should look like.

My challenge was to shape EMILY's List into an organization that would appeal to our prospective members. Everything — our name, the way we operated, the way we communicated, our corporate culture — had to show prospective members that we were an organization they would want to be a part of and to help succeed.

Just as we didn't call EMILY's List something like the Democratic Women's Campaign Fund, or another one of Washington's unimaginative mouthfuls, we designed a creative letterhead — a kind of takeoff of the Fleischmann's yeast logo — and sent out a personalized message to people we thought could give a little more than the average. The whole idea was to distinguish us from other women's groups.

AT THIS EARLY STAGE OF 1985, when it came to the idea of electing a woman to the Senate, it was still not entirely clear who would be running. At various times, the rumor mill had it that former first lady Rosalynn Carter might run for the Senate from Georgia, that Bethine Church, the widow of Sen. Frank Church, might run in Idaho, and that astronaut Sally Ride might run in California. But none of them did.

That left us with just two candidates, both of whom were terrific contenders. In Missouri, Harriett Woods had just been elected lieutenant governor and was running again for the Senate in '86. Her loss in 1982 had been a real wake-up call for us, and now, armed with statewide name recognition and a national donor base, she was in a perfect position to avenge that defeat. The other candidate was Barbara Mikulski, who was serving her fifth term as a congresswoman from Baltimore and had already begun the process of preparing to run for the seat occupied by Republican senator Charles Mathias.

The Catholic daughter of a Polish grocer, Mikulski was one of a kind. As a student at Mount Saint Agnes College in Baltimore, she had seriously considered joining a religious order before her religious ideals collided with her relentlessly rebellious nature. "Poverty was one thing and I could go along with chastity," Mikulski told a reporter. "But it was obedience. I thought, 'My God, all my life there could be someone telling me what to do and where to go!' And inside me beats the heart of a protester!"

So, instead of becoming a nun, Mikulski became a social worker. In the sixties, she registered blacks to vote, organized tenant strikes, and followed the teachings of visionary idealists such as Dorothy Day, founder of the Catholic Worker movement; Msgr. Geno Baroni, who fought to preserve Baltimore's ethnic neighborhoods; and Saul Alinsky, the legendary Chicago community organizer. Described as "a stocky, 4-foot-11, rough-edged East Baltimore politician," Mikulski first won national attention when she wrote an article in the *New York Times* in September 1970 asserting that ethnic America was "forgotten and forlorn" and that "elitist" liberals unfairly branded working-class ethnics as racist, Archie Bunker–like stereotypes. In the mid-seventies, she successfully led community opposition to a sixteen-lane highway that would have ravaged the Fells Point neighborhood and Baltimore's Inner Harbor, both of which remain thriving communities today.

In winning a local battle that got national attention, Mikulski proved she was a natural-born pol and earned a reputation as a feisty, street-fighting urban populist who fought the powerful city hall machine and big-money developers and won. In doing so, she adroitly bridged the chasm that often separated the traditionally conservative, white working-class constituency she grew up with and the more elite, progressive wing of the party.

Community organizing has been an unusual spawning ground for high-level Washington politicians, but, as Mikulski showed, in many ways it was a natural starting point for women. "A lot of the early-bird women in politics got in because of activism," she said. "For one reason or another, we all ran into huge political opposition, people who kept telling us, 'No, it's not our thing. What you want is not our priority.' And eventually, after hearing those 'nos' too many times, we developed a motto: 'Don't get mad. Get elected.'"

So, for Mikulski and other future senators, what counted initially was having a base in the community, an ability to marshal volunteers and to put together a grassroots coalition.* "The next big step was to run for federal office," said Mikulski, "and that usually meant a congressional district where we already had a base." So, in the seventies, we had a cornucopia of women coming to the House, and they had names like Bella Abzug, Pat Schroeder, Barbara Jordan, and Shirley Chisholm.

Mikulski had been charting a path toward higher office since 1974, when, as a member of the Baltimore City Council, she ran for the Senate against Charles Mathias, a popular liberal Republican incumbent. With little statewide name recognition and only $40,000 in her war chest, Mikulski ran a first senatorial campaign that proved quixotic.

---

* Barbara Boxer, who was later elected United States senator from California, was first elected to the Marin County Board of Supervisors in 1976. Dianne Feinstein, who later became the senior senator from California, started out on the San Francisco Board of Supervisors. Patty Murray and Maria Cantwell, both of whom became senators from the state of Washington, got their starts, respectively, opening day care for children and keeping the public library open.

Even though she lost, she was unfazed, having astounded observers by winning 43 percent of the vote and positioning herself for a congressional seat. After her defeat, she was asked whether she considered her political career dead. "Are you kidding?" she replied. "I'm a household word — like Brillo or Borax."

Two years later, Rep. Paul Sarbanes gave up his seat in Maryland's Third Congressional District to run, successfully, for the Senate. Mikulski took his place and never looked back.

The first woman to serve on the powerful House Energy and Commerce Committee, Barbara was startlingly adept at penetrating the all-male, bourbon-sipping culture that ruled Capitol Hill. "I didn't want to be on the pink-collar ghetto committees," she said. "Energy and Commerce has the big boys, the chieftains." So, Mikulski made a point of wooing the politically canny Tip O'Neill, who was soon to become Speaker of the House, until she finally broke the gender barrier.

A stalwart supporter of labor and women's issues, Mikulski soon became known as one of the toughest and most effective Democrats in the House. In mid-May 1983, the party leadership selected her to give the Democratic Party's response to President Reagan's weekly radio address. After being touted as a potential running mate for Mondale in 1984, she became a fervent supporter of Ferraro, whom she coached for her vice presidential debate with George H. W. Bush. Moreover, as a leader in the women's movement, an important adviser to Mondale, and a spokeswoman who addressed the Democratic National Convention three times in four days, she had begun to make a name as a national figure.

Then, in 1984, while she was preparing to run for her fifth term in Congress, Mikulski became cochair of the Mondale-Ferraro campaign, fully aware that her position would provide her with important connections if she wanted to move up the political ladder. "We all yearn for the next domain," she said. "So, I put together a strategy group of women from both Baltimore, including my chief of staff,

Wendy Sherman,* and people I met on the Mondale campaign." For most of her electoral career, when it came to fund-raising, Mikulski had eked by with nickel-and-dime political contributions. "When I began thinking about the Senate, my friend Ann Lewis said, 'Barbara, you've got to think beyond the bake sale.'" The Mondale-Ferraro campaign gave Mikulski an opening to woo donors outside her congressional district. "This was before the era of PACs, and there was no such thing as a Bette Midler benefit, where you'd raise a million dollars," said Wendy Sherman. "So, I wrote a memo of what Barbara should do if she wanted to run for Senate." The bottom line: Mikulski had to get to know the big donors.

But, in 1985, as soon as she prepared to run, Mikulski immediately encountered two very formidable obstacles: Harry Hughes and Mike Barnes. The former was a popular sitting governor with term limits who was looking for another job, a solid man with dashing good looks. The latter was one of Mikulski's fellow congressmen, a rising star from Montgomery County, Maryland, who was particularly adept in foreign policy and an expert on Latin America at a time when that expertise was highly relevant. "When we saw the primary field — a successful sitting governor and a popular member of Congress — we knew it was going to be a tough primary battle," said Wendy Sherman.

At the time, about two out of three voters in Maryland were Democratic, so, if Republican Charles Mathias declined to run for his Senate seat again, the Democratic primary was likely to determine the state's next senator. And, as Mikulski's team saw things, having enough money for that primary — early money — was the prerequisite for the entire election. "The first Federal Election Commission report *had* to show that we were competitive," said Wendy Sherman.

---

* I later hired Wendy Sherman to be executive director of EMILY's List. Still later, she became undersecretary of state for political affairs in the Obama administration, and was one of the chief negotiators in the nuclear negotiations with Iran in 2013 and 2014.

Thanks to mutual friends, Mikulski had been aware of EMILY's List from the word go. "I just loved the name — Early Money Is Like Yeast," she told me. "I loved it because my grandmother ran the best Polish bakery in Baltimore, and when I was running for city council, a woman who had the Baltimore beehive hairdo that got famous in the movie *Hairspray* asked if I was any relation to Mikulski's bakery. I told her I was, and she said, 'Listen, kid, if you're half as good as your donuts, you'll be okay.' So, when I heard what EMILY stood for, I thought, 'This is my crowd.'"

IN LATE JANUARY 1985, I got a call from Wendy Sherman asking whether I'd have lunch on the Hill with Representative Mikulski. I was a nervous wreck. I was so green that I'd never even met a member of Congress before.

After we sat down in the House dining room, Mikulski explained her strategy. She was more than viable, she said, because she was enormously popular in Baltimore — she had just been reelected by a two-to-one margin — and to win the entire state, a candidate needed Baltimore. Mike Barnes was barely known there.

Then, it was my turn. I explained that EMILY's List could raise much more than the $5,000 PAC limit because of the way we were set up, and if one thousand people each wrote a check for $100, we could raise $100,000. It was pretty much the same rote speech I gave to women whom I hoped would join EMILY's List as members. But this situation was different — very different.

Mikulski turned to Wendy Sherman. "That's great," she said. "I can't wait!"

Then, she took out a piece of paper and started writing. "I've got you down for a hundred grand," she said.

It was all so simple, but I just sat there, trembling.

"Holy smokes!" I thought. "Now, I really have to do this." And I had absolutely no idea whether we could really deliver.

Seven

A DISTANT KINGDOM

COMING UP WITH THE IDEA for EMILY's List was one thing. But now we had to show that our ideas really worked.

From the beginning, we took pains to create a structure that would allow us to avoid the heated organizational strife we had seen at NOW and other feminist groups. As for the women I'd brought together, we liked each other. There was genuine camaraderie. So far, there had been no real internal political battles or issues of trust —and we wanted to keep it that way. We decided to avoid having bylaws and a highly visible board of directors; we knew that people would lobby those directors to make us endorse particular candidates. Instead of going through that craziness, we appointed a steering committee including Betsy Crone, Joanne Howes, Marie Bass, and Judy Lichtman, who would meet only when necessary. Period. It proved to be a very popular move that insulated us from internecine political battles.

We also had to nail down the criteria a woman candidate had to meet if we were to endorse her. Good-government me thought that maybe we needed to canvass women like representatives Mary Rose Oakar (D-OH) and Lindy Boggs (D-LA), two highly regarded but anti-choice women, as to whether the pro-choice stance of EMILY's List would preclude their participating. To that end, I met with Oa-

kar and Boggs, and then reconvened with the steering committee
to debate the issue. But before the discussion got very far, someone muttered, "Of course, we'd only want to help women who are
pro-choice."

It was one of those spontaneous comments that produced a moment of striking clarity. Everyone agreed—so readily that we immediately stopped our debate. We knew what we believed in and
what we wanted to fight for. It made no sense to say that women
are equal participants in America today but, by the way, they don't
get to control their bodies. Reproductive freedom and the ability to
control our own bodies were, in our minds, fundamental to women
having an equal place in our society. This is when we decided we
would endorse only women who were running as Democrats and
who were pro-choice.

KNOWING THAT THE BEST way to find our market segment
was to reach out to like-minded women, we decided to invite our
friends to my house on Oregon Avenue in northwest Washington to
address solicitation letters. In that long-ago predigital era, most of
us stored our treasured collections of addresses and phone numbers
on Rolodexes, rotating file devices that are an anachronism now but
were then ubiquitous in offices all over the country. We called these
events "Rolodex parties."

Our first Rolodex party took place in April in the basement at
my home with about twenty-five political friends. We all sat at long
tables, and, over coffee, we told our friends all about our plans for
EMILY's List and asked them to join and to write and invite their
friends as well. Then, we all went through our address books and
Rolodexes, sharing information about each other's lives as we addressed envelopes.

As the night wore on, one person after another shouted out a
name. "Is anyone doing Billie Bobbitt?" (Bobbitt was a woman U.S.

Air Force colonel who was a leader in the Florida ratification of the ERA.) Across the room someone would yell, "Oh, I want to send it to her!" So, both women agreed to write a note on Billie's letter. By the end of the night, we had addressed almost 450 letters. When they were sent out and the returns came in, we ended up with more than one hundred members.

A week or so after the letters went out, I remember standing with Betsy Crone at the front desk at 2000 P Street, opening the mail just after it had arrived. After going through a handful of letters with checks from friends and relatives, Betsy opened an envelope and looked quizzically at the signature on the check.

"Ellen, do you know who this is?" she asked.

"No, I've never heard of her."

We both smiled. We had received our first contribution from someone we didn't know. We realized what it meant: some complete stranger thought what we were doing was such a good idea that she actually sent in a check.

Or, to put it another way, EMILY now had a list!

IN JUNE 1985, EMILY'S LIST had 108 members. That was good enough to warrant another Rolodex party on July 17 — this time on a larger scale. Again, we asked our friends to write a short note on our printed letter, to address their envelopes, and to seal the letters inside. "These early meetings were terrifically important in that they emboldened us," said Judy Lichtman. "They animated us. This was a time when reproductive rights were really coming under attack from Republicans."

Over the next year, I took Emily's List's show on the road, to Atlanta, Boston, Chicago, Denver, New York, Philadelphia, San Francisco, and more. In one town after another, we replicated our Rolodex parties. I asked women in the audience to contribute $100 to be a member. The essence of my pitch was that EMILY's List would be an effective organization that would help elect Democratic women to office.*

Initially, I was a bundle of nerves, because I wasn't used to public speaking. But over time, I got better. I talked to John Gardner, who had founded Common Cause, and he told me he was so petrified of speaking in public that his hands shook. But he finally realized that others didn't know, so it didn't matter. I finally reconciled myself to the notion that no matter how awful I felt, I wasn't going to pass out. I absolutely believed to my core that what we were doing made sense. And when I said that if one thousand women wrote checks for $100, we could raise $100,000 for a woman candidate, sometimes people actually started cheering. They were getting excited — and that made it much easier for me.

I refined my approach over time. I soon realized that the fervor faded fast for even the most enthusiastic people, so I made a point of standing by the door to collect $100 checks *before* they left. Soon, I was signing up as many as ten or fifteen new people each time. By the fall, we had more than five hundred members. EMILY's List was growing.

---

* A typical pitch went like this:

EMILY's List is a network of men and women who want to have more progressive, pro-choice, Democratic women in office. EMILY stands for Early Money Is Like Yeast. We raise early money for our candidates to build their credibility, and we continue to support them until they win.

Here's how we work. We recommend to you viable pro-choice, Democratic candidates who have a realistic chance of winning. We tell you about their campaigns and their positions on issues. And then you decide who you want to support. You write your checks directly to the candidates and send them back to us. We combine your checks with hundreds of others and send them directly to the campaign.

Under the current law, if EMILY's List contributed directly to the campaign, we could only give up to $5,000 in the primary and general election. But if one thousand EMILY's List members write checks made out to the Mikulski Campaign for $100 each, we can raise $100,000 as opposed to the $5,000 or $10,000 we could otherwise contribute. That's how we make a tremendous difference. That's how we make history!

It's $100 to be a member of EMILY's List, and we hope you will write checks to two or three candidates we recommend during the election cycle for $100 or more each. So I hope you will join EMILY's List right now! Let's all work together to make history. Let's elect the first Democratic woman senator in her own right!

Before long, we began to make a name for ourselves. Our Rolodex parties were almost chic. In Washington, Anne Bingaman, the wife of Sen. Jeff Bingaman (D-NM), hosted a Rolodex party for the wives of senators. Walter Mondale's wife, Joan Mondale, mailed two hundred letters after one party. In Minneapolis, one EMILY's List member referred to our gatherings as "the ultimate Tupperware party."

In Miami, Houston, and elsewhere, again and again I recited my mantra about early money. Together, the mailings and our presentations began to have a real impact. "At first, I was intrigued by the name," said Sherry Merfish, a Houston lawyer and activist who received the first direct-mail solicitation, in 1986.

> I remember reading it, and I immediately thought this was a viable solution to the problem of the underrepresentation of women, that my contribution alone might be insignificant, but if it was bundled, we could really make a difference. I immediately recognized the genius of it.
>
> I had already been persuaded by the mailing. But when Ellen came to Houston and I went to dinner and met her, I suddenly became zealous about it. Her passion and the fact that she was so pragmatic about it bowled me over.

Thanks to women like Sherry, we soon reached our initial goal of one thousand members. We were beginning to make ourselves known. In June, we issued our first press release, announcing the formation of EMILY's List, for which we were rewarded with short news stories in the Associated Press and the *New York Times*.

DURING THE SUMMER OF 1985, Maryland voters faced an embarrassment of riches as three promising hopefuls — Mikulski, Gov. Harry Hughes, and Rep. Mike Barnes — began their campaigns for the Democratic senatorial nomination in Maryland. "We have three of the best candidates," said one elderly voter. "If only we could spread them around the United States instead of having to choose between them."

In August, however, Hughes encountered a problem that was not truly of his own making. The savings-and-loan scandal sweeping the nation was erupting in Maryland, with about 137,000 depositors being unable to withdraw money from four institutions in the state. Because it was happening on Hughes's watch, he soon became known as "the incredible shrinking governor."

By mid-September, Barnes, seeing that Hughes had lost his luster, sought backing from the party leaders, as Mikulski worked quietly behind the scenes. Two weeks later, as had long been expected, Senator Mathias formally announced his retirement, making it all but certain that the winner of the Democratic primary would be the next senator from Maryland. Even though Mikulski had not formally announced, at about the same time a *Baltimore Sun* poll showed her with 36 percent of the Democratic vote, Hughes with 18 percent, and Barnes at 16 percent.

Both Harry Hughes and Mike Barnes were good, solid liberal Democrats, and for the first time we at EMILY's List realized that progressive male adversaries presented a potential conflict for us. As it happened, the former director of Barnes's district office was none other than Marie Bass, one of our Founding Mothers. In other words, one of our own was a key figure in the kitchen cabinet of a political foe. But, rather than let an imminent conflict fester, Marie swiftly addressed the issue head-on, with style and grace. "I love Mike Barnes," she said. "I think he'll make a fabulous senator. But there's no doubt in my mind what EMILY's List should do—and that is, we have to get behind Barbara Mikulski."

Marie's aplomb helped shape the collegial tone with which EMILY's List functioned in the contentious and often bitter world of Washington politics. Ultimately, like everyone in Washington, each of us had her own priorities. But at EMILY's List, we learned to check our personal agendas at the door.

★   ★   ★

ON MONDAY, OCTOBER 28, 1985, Barbara Mikulski finally announced her candidacy. "I want to be the senator for all the people of Maryland — the white collar, blue collar, pink collar, new collar, and no collar," she told a congressional committee hearing room packed with prominent feminists. She noted that the women were there "in support of sisterhood and solidarity. They were there to see astronaut Sally Ride take off. Now they're here to see Barbara Mikulski take off." If elected, the *Washington Post* noted, she would be the first woman to be elected senator from Maryland.

With EMILY's List's help, Barbara was putting together all the key elements necessary to ascend the political ladder. "The Senate was like a distant kingdom," Barbara later said. "That's because to do a statewide race, you needed a really significant infrastructure and the money to be able to pay for it and statewide media. I often said that every woman who runs for office needs a 'MOM.' You need a Message, you need an Organization to get out your message, and you need Money to pay for the organization."

Even with EMILY's List behind her, Mikulski still faced a difficult unspoken battle. To the Democratic Party in Maryland, Mike Barnes looked far more like a senator than she did. "What I was fighting most was not them so much as the stereotypes of a woman candidate, particularly someone who is short, chunky, and mouthy," she told me. "I'm not particularly glamorous-looking, and it was the idea that I just didn't look the part." All of which Mikulski confronted with a typically disarming response. "I guess togas don't come in a size 14 petite," she said.

Indeed, it was precisely her spontaneity and authenticity that became Barbara's great selling points. Far from being a plastic TV personality, she was one of the people. "That's what the voters wanted," Mikulski said. "They wanted somebody who would be a fighter, somebody who looked like them, fought like them, talked like them, and would stand up for them. And that's what I showed them." So, when she lost weight and people slapped her on the back and told her she was looking good, she always had a great comeback. "Yep, I'm counting my calories, I'm counting my votes, and I'm counting on you on Election Day," she said.

Being totally natural and authentic meant Mikulski really *was* out of place in the Senate, and she played it up to the hilt. "There are some senators who eat quiche," she said. "I eat bacon and eggs. They told me I needed to wear these fancy-girl shoes called Ferragamos" — sometimes she called them "Feggaramos" — "and I thought it was a restaurant in Baltimore. All I know is that they cost a hundred bucks and they pinch my feet like hell."

The voters loved it — and polls showed as much. In November 1985, a WBAL-TV survey showed that Barbara had widened her already substantial lead. The numbers looked great.

The only problem was that the numbers had looked good for Harriett Woods four years earlier, and that hadn't been enough. And there was another set of numbers that were deeply, deeply troubling. On December 31, the candidates filed their year-end federal election fund-raising reports. Rep. Mike Barnes had raised $349,950, while Barbara had raised only $197,000. Worse, Barnes had more than twice as much money on hand as Mikulski. "In spite of all the talk surrounding Mikulski's early lead, we find that Mike is at a very strong position financially," said Bill Bronrott, spokesman for Barnes.

In other words, Barbara Mikulski was nationally known, had a sky-high approval rating, and a 2–1 statewide margin in the polls, but she still had not won the backing of the Democratic Party establishment in Maryland. It was astounding. The failure of the state establishment to support women was exactly what had defeated Harriett Woods four years earlier. This time, the numbers in Mikulski's favor were even more dramatic. The situation represented everything EMILY's List had been created to fight.

IN JANUARY 1986, after nearly a year building an organization and developing a membership, it was finally time for EMILY's List to raise money for our candidates. At last, we would see whether or not our ideas really worked.

To do that, first we had to create profiles of candidates that would go out with mail solicitations. We started by inviting women who had expertise in various fields to help us assemble questionnaires surveying the candidates on a dozen or more issues, from the budget to nuclear disarmament, foreign policy to unemployment, and more. Mikulski and Woods responded with thoughtful answers that we used as a basis for their profiles. Because EMILY's List was not a lobbying organization, we weren't trying to promote specific positions. Instead, we simply wanted our members to understand the positions of the candidates we were endorsing.

In February, we included those profiles in the first "candidates mailing" to our membership list and asked members who had already contributed $100 to EMILY's List to now donate $100 each to Mikulski and Harriett Woods. We simply did not know whether they would.

So, the waiting began.

Fortunately, we didn't have to wait long. Checks poured into EMILY's List almost immediately. And they were not just for $100 but for $250 and $1,000, split almost equally between Mikulski and Woods. In just three weeks, we raised $22,986 for Woods and $23,916 for Mikulski. That was more than four times what an ordinary $5,000 contribution from most PACs would have been. Our donor-network idea actually worked, and we were elated.

Barbara had expected her money to come in one fell swoop, but instead it came in gradually. "They were like little bundles of joy," Mikulski said. "Every week, Wendy, who had become my campaign manager, would get the mail and start opening the envelopes, saying, 'Little EMILY's arrived!'"

Meanwhile, having started out strong on the campaign trail, Barbara simply got better. According to the *Washington Post*, her "quick quotable wit, strong stands and hard work" were evident thanks to "a rousing rhetorical style marked by an outrage that seems as fresh as it was in 1970." Her once imposing Democratic foes were no longer so frightening. Still dogged by the savings-and-loan scandal, Governor Hughes, who had been elected with 71 percent of the vote just a few years earlier, was effectively finished. "It's the death by a thousand cuts," said one Hughes supporter.

In Hughes's place, Mike Barnes had become the darling of the Maryland political establishment. At the same time, even though he had been winning headlines for opposing President Reagan's efforts to fund the contras in Nicaragua, nothing he did helped him gain traction.

In early March, the money continued to flow into our office. The quarterly Federal Election Commission filings were due to be reported on March 31. It was a long-awaited moment for us, because it represented an objective bottom-line assessment of whether Barbara was really competitive in fund-raising.

When the numbers came in, we were astounded: Barbara had raised a total of $306,642 that quarter, while Barnes had raised only $186,206 and Hughes fell just shy of $100,000. "A year ago, opponents said she couldn't win votes, but the polls showed she could," announced Wendy Sherman as Mikulski's campaign manager. "Last fall, opponents said she couldn't raise the money, but this quarter showed she could."

And what role were we playing? EMILY's List members had contributed $60,000 of that total to Barbara. "EMILY's List is providing the margin of victory!" said Mikulski.

Now, for the first time, we saw the power of early money working *for* a woman candidate, rather than against her. The *Washington Post* referred to Mikulski as the "front-runner" with a "commanding lead" in the race to replace Mathias.

Barnes courted black voters in Baltimore by attempting to capitalize on lingering hostility between the black community and the white ethnic communities tied to Mikulski. But it didn't work. Less than a week later, Barbara received the backing of the twenty-six churches in the Baltimore AME Ministerial Alliance, not to mention that of black business leaders, a black city councillor, and a black Democratic club.

By May, some four months before the primary was to take place, her lead was so unassailable that the *New York Times* asserted — erroneously, of course — that Barbara was *already* the Democratic nominee. A month later, a *Washington Post* poll gave Mikulski a huge 28-point lead over her rivals. Campaign-finance reports showed she had far more cash on hand than her two foes combined.

MEANWHILE, IN MISSOURI, Harriett Woods, our other candidate, faced a very different kind of race. As a sitting lieutenant governor seeking the United States Senate, she did not have serious challenges in the Democratic primary, as Mikulski did. Her big battle was going to be the general election in November, in which she hoped to fill the seat being vacated by retiring senator Tom Eagleton. Her main problem was that the GOP had barely beaten her in 1982, when she nearly upset incumbent senator John Danforth, and the Republicans were not going to make the mistake of underestimating her again. As a result, the Republican Party was unstinting in its support of Republican nominee Christopher "Kit" Bond, a good-looking, broad-shouldered, wealthy, sixth-generation Missourian and former governor of the state. Huge amounts of money came from out of state for Bond, and both President Reagan and Vice President George H. W. Bush made appearances on his behalf.

Through most of June, the campaign was neck and neck. Woods and Bond were both attractive, well-schooled, thoroughly professional candidates who opposed each other on virtually every major domestic and foreign-policy issue. According to one poll after another, each had roughly half of the electorate. And this time, unlike in 1982, Harriett was finally getting support from the party regulars. Between them and EMILY's List, she had amassed more than $1.2 million in campaign funds by June, enough to keep her in the running and just $200,000 behind Bond's growing war chest.

Then, in late June, came one of those moments in politics that's like a train wreck you see coming but can't do anything about.

It's a rule of thumb that the vast majority of TV viewers forget political ads moments after viewing them. But every so often, an ad comes along that sticks in the public consciousness and has an enormous impact. In 1964, Lyndon Johnson's famous "daisy commercial" was one, showing a two-year-old girl in a field innocently plucking petals off a daisy while, in the background, a countdown leads to the launch of a nuclear missile followed by a billowing mushroom cloud. The unspoken but powerful message — that GOP nominee Barry Goldwater was a dangerous warmonger — helped Johnson win by an overwhelming landslide.

According to Don Sipple, Kit Bond's media consultant, Harriett Woods's "crying farmer ad" had an equally dominant role in this campaign — only this time it backfired on the Democrats. The ad, which was put together by legendary Democratic media consultant Bob Squier, focused on David and Marilyn Peterson of DeKalb County, Missouri, whose farm had been foreclosed by Mutual Benefit Life, an insurance company of which Kit Bond was a director. Woods, who had been making a name for herself as a champion of the little guy, had proposed a moratorium on farm foreclosures, and the idea was to portray Bond as a cold-hearted, latter-day Simon Legree who didn't think twice about letting banks take over the sacred family farm.

According to the book *Madam President,* by Eleanor Clift and Tom Brazaitis, after seeing the ads for the first time, Woods told Squier, "Bob, you may win an award for these, but I could lose an election."

"Don't worry," Squier replied. "This is a silver bullet. I won't use this until the fall, or when we really need it. This is a killer. We'll put him away with this one."

By July, Bond had begun to rack up a big lead in fund-raising, and Squier decided it was time to strike back with his "silver bullet." But when the ads actually aired, the sight of David Peterson in tears elicited a response that quickly spun out of control — in the wrong direction. All over the state, press coverage was merciless: "Mindless emotionalism" . . . "What bunk! What tripe!" . . . "Weepy, wimpy TV extravaganza" . . . "crass demagoguery."

In August, I flew down to Missouri to see Harriett's operation firsthand, and I was impressed no end by the whirlwind campaign she was running. After meeting her at 7:30 a.m. at the Saint Louis airport, we flew 175 miles to a toxic-waste site in Springfield, Missouri, to make sure Harriett got a good hit on the evening news. Then, in rapid succession, Harriett addressed her supporters at a lunch in Springfield; flew to the Ozarks to meet members of the United Auto Workers; invaded the Republican heartland in Waynesville, Missouri; visited two radio stations; and drove on to address rallies in Rolla and Steelville, Missouri — all the while giving interviews to journalists in the car along the way. And this was all in one day!

Harriett's indefatigability notwithstanding, the damage from the "crying farmer ad" was both deep and lasting. In Columbia, Missouri, the *Daily Tribune* published a story saying that the farm family depicted in the ad, the Petersons, felt exploited by the Woods campaign. More than a month after the ad series was broadcast, an unidentified voter who said he was politically independent approached Kit Bond. "I don't know who I'm going to vote for," he said, "but I just want to tell you I thought those ads Harriett Woods ran against you were the most offensive I've ever seen."

Even Democrats jumped on the bandwagon against her. "She shouldn't show a man crying in public," said state senator Truman Wilson, a Democrat. "She should have had a Marlboro man, looking steadily into the distance, bearing up under adversity."

Harriett fired Squier, but the damage was done. By September, Bond was 8 points ahead in the polls.

I ALSO HAD A CHANCE to spend time in Baltimore and got to see how different Barbara Mikulski was from the run-of-the-mill politician. Where most politicians could be intimidating and off-

putting, Barbara had wit, energy, good humor, and a genuine, disarming warmth that put people at ease, that allowed them to approach her almost as if she were a member of their family.

She was so good that in August, Sen. Edward Kennedy (D-MA)
broke with protocol—he had never before endorsed a candidate
during a Democratic primary—and threw his support behind Barbara. "I have decided to make this endorsement because I regard
Barbara Mikulski as a special person, a unique candidate, who will
be a unique and powerful presence in the Senate," Kennedy said.

With just three weeks left until the September 9 primary, Mike
Barnes's funds were so low he was forced to fire his pollster and his
media consultants, and to cut the pay of his campaign staff. Barbara's position was so strong that the *Washington Post* declared her
"untouchable" and quoted Democratic consultant Peter Hart as saying that Barnes's only chance was to hope "for a real natural disaster in Baltimore that doesn't strike the Washington area." Pundits
now turned their attention to what would likely be an all-woman
general election, with Mikulski running up against Linda Chavez,
a conservative Republican who had campaigned vigorously against
affirmative action, and who was the probable GOP nominee.

Then, just before the primary, I got a call from Anne Wexler, asking whether I needed a ride to Mikulski headquarters to watch the
election results for the primary. I had to smile. Anne Wexler wasn't
just anybody. A longtime savvy political operative, she had worked
for George McGovern's presidential campaign in 1972 and was responsible for discovering and hiring a young graduate of Yale Law
School named Hillary Rodham to her first job in politics, heading
up a voter-registration drive in Texas. Anne's husband, Joe Duffy,
had been chairman of Americans for Democratic Action, but Anne
herself had gone on to become a high-ranking troubleshooter in
Jimmy Carter's White House and was often cited as one of the first
women to become a real White House insider. "It depends how tight
you drew the circle," said Jody Powell, Carter's former spokesman.
"If it was three or four—no. But if it was five or six—yes, Anne was

absolutely there." Then, in 1984, Anne had become a senior adviser to Geraldine Ferraro. Now she was close to Barbara Mikulski.

In other words, Anne was one of the very first women to break through into the old boys' inner sanctums of power. She was not just a player but a gatekeeper, and now she was inviting me in, when not long before I had been an innocent naïf anxious about meeting Barbara Mikulski in the House dining room for lunch.

I had never even been to an election-night event before. Stepping out into the crowded hotel ballroom, with hundreds of raucous people, made it clear to me that the word was out, that EMILY's List was doing something significant. When all was said and done, Barbara got 343,432 votes, or 53 percent, in an eight-candidate field, leaving Mike Barnes in a distant second place with 189,531 votes, or 29 percent. Smiling broadly, in her acceptance speech Barbara said her Democratic opponents had pledged to "make sure the Mikulski victory is not a one-night stand," a reference to her upcoming general-election battle against Republican Linda Chavez. "This is a victory that belongs to everyone," she declared. "The people were with Mikulski, and Mikulski is with the people."

NOW THAT THE PRIMARY was over, Mikulski faced a new foe. Linda Chavez, as director of public liaison, was the highest-ranking woman in the Reagan White House. Because Maryland was strongly Democratic, Barbara was the clear favorite, but we weren't counting our chickens yet.

Before we knew it, Chavez had launched a full-bore campaign replete with gay-baiting characterizations of Mikulski as a "San Francisco-style liberal" who was "anti-male." To which Barbara replied calmly that this would come as "a real shock to my dad, my nephews, and the guys at Bethlehem Steel."

But Chavez's problem was not merely Barbara's snappy rejoinders. More to the point, Mikulski had cobbled together a powerful and cohesive assemblage of middle-class Jews, blacks, and polyglot working-class ethnic voters in a way that could have been done only

by an old-fashioned Democrat who was a working-class ethnic herself. Barbara's voice rang true because she had been part of those
communities her entire life. When she went out on the hustings,
people reached out to her as if they knew her and understood that
she would fight for them. So, when Chavez characterized Barbara
as an abrasive, far-left fringe character, no one bought it. They already knew her. "She spent fifteen years in the dominant market in
the state, building a base that can't be touched," said Democratic
Party treasurer James Rosapepe. "Even a well-run Chavez campaign
would have gotten clobbered."

Even before the general election, Barbara began to jockey for a
suitably prominent place in the Senate. Having already served ten
years in the House, she was not exactly a neophyte and was hardly
the type to arrive in the Senate dazed and starry-eyed. Early on, Barbara met with Paul Sarbanes, Maryland's other senator and a fellow
Democrat, who was gracious enough to become an early mentor.
She cultivated relationships with senators Robert Byrd (D-WV),
Teddy Kennedy (D-MA), and Howard Metzenbaum (D-OH).

Normally, freshman senators were required to do time on one of
the less prestigious committees. But, even before November, Barbara began seeking a seat on the powerful Senate Appropriations
Committee, which oversees almost every aspect of government
spending. Maryland was home to hundreds of thousands of federal
civilian employees and government retirees, so she was eager to get
a position that gave her maximum input on civil-service matters.
Plus, the Appropriations Committee was where the money was.
Barbara Mikulski entered the Senate ready to leverage her win.

At the same time, Barbara made a point of reaching out to me to
make sure that Harriett Woods also positioned herself to get on a
powerful committee if she won. It was a sisterly gesture and a pre-
cursor to her role as dean of women in the Senate. More than that, it
meant she was grateful for the role EMILY's List played in her campaign and saw the organization as a potentially powerful vehicle to
elect women to Congress. Again and again, she said that she would

be happy to be the first woman Democrat elected to the Senate in her own right — but she also wanted to be the first of many.

Harriett Woods, unlike Barbara, hadn't had the burden of a tough primary battle, but the general election was always a challenge in the Republican-leaning state of Missouri. As the November elections approached, Woods trailed Kit Bond by a handful of points. She did all she could do, but going into Election Day, November 4, it looked as if it just wouldn't be enough. So, that night, when Harriett lost, I was disappointed but not terribly surprised.

The bigger news that first Tuesday in November was also expected, but it was historic nonetheless. In Baltimore, just a few blocks from where she began her battle to defeat the expressway, Barbara Mikulski reveled in the power of neighborhood politics that enabled the scrappy daughter of a Polish grocer to rise to the august chambers of the United States Senate, the first Democratic woman elected in her own right! EMILY's List had made history, too, in its very first election, and we were elated.

Four years earlier, Harriett Woods had gone down to defeat because the old boys in the Democratic Party wouldn't support her. Similarly, this time, even though Barbara started with a 20-point lead in the polls, the party establishment believed that Mike Barnes was the only figure in the race who was "senatorial" enough. The difference was that EMILY's List had entered the picture. After the first quarterly Federal Election Commission reports were filed, the old boys saw that Mikulski actually had more money than Barnes. Mikulski's victory proved that our early-money strategy was the right way to go. We had started a revolution by checkbook. It had the potential to ignite the ascent of Democratic women in politics all over the country.

When we had launched EMILY's List ten months earlier, we started off with three clearly defined goals. First, we wanted to have at least 1,000 members. We already had 1,200. Second, we wanted our members to contribute at least $100,000 each to Mikulski and Woods. Again, we exceeded our goal, with our members giving a total of $350,000 for the two of them. When we included what we raised to administer EMILY's List, the cumulative total was nearly $500,000. Finally, we wanted to make history by making sure the first woman Democrat was elected to the Senate in her own right. Now that all the votes were counted, Barbara had won a smashing 22-point victory over Linda Chavez.

We had achieved each of our three goals by such a wide margin that when I thought about it, I told myself, somewhat facetiously, that perhaps we had made one mistake after all: maybe we had set our sights too low. The next step was to aim higher, to make sure that we were not a one-shot wonder, to build EMILY's List into an institution that could grow and endure and really reshape the role of women in American electoral politics. That meant we had to find and encourage women leaders who could play major roles on the national stage. And we had to reverse the Democratic Party's skepticism toward women.

Eight

A STAR IS BORN

HAVING SECURED HER coveted seat on the Appropriations Committee, Barbara Mikulski made it clear that she knew how to play the game. But that was merely the beginning. Many senators had expected Barbara to be, as the *Washington Post* put it, "the second coming of Bella Abzug" — that is, a brassy, bomb-throwing feminist who would make a lot of noise but not really get things done.

They were wrong.

One of the earliest tests in Barbara's tenure took place less than a month after she took office, at a banquet room in a Washington hotel where 650 senators, business leaders, cabinet secretaries, and Supreme Court justices joined Vice President George H. W. Bush for the Alfalfa Club, an all-male Washington social club that exists only to hold an annual banquet each January involving silly skits, jokes, musical performances, and the like.* Late in the evening, Sen. Pete Domenici (R-NM) delivered an impromptu satirical campaign speech that was typical of the high jinks that went on at this event. "I'm blessed with the talent of . . . whipping the electorate to a frenzy," he proclaimed. "Just like the singer Tom Jones, women

---

* The Alfalfa Club began admitting women in 1994, twenty years after it began admitting blacks.

often throw their panties at me when I speak. It happened again just yesterday. I just don't know what got into Senator Mikulski."

The next afternoon, when Mikulski heard about the remark, she realized she was facing a defining moment in terms of her image as a senator. "I was livid," she said. "Livid."

Then, she surprised some observers by responding with a two-tiered strategy. First, she issued a public statement. "I think it's outrageous and I find it insulting," she said. "I have other responses, but that's the one for public dissemination." Privately, however, Barbara used the occasion to shore up support on the Senate floor. When a contrite Domenici called the next morning to apologize, Mikulski, according to the *Washington Post,* helped him draft an apology, was conciliatory, and said she considered the matter closed. Two months later, after he wrote an article on the homeless, Mikulski seized the occasion to praise him on the Senate floor.

She was putting out a message loud and clear: having made her name in Baltimore as a street-fighting community activist, Mikulski had become senatorial. "I find her style more relaxed, good-natured, a little less shrill and harsh," said Linda Dorian, executive director of Business and Professional Women/USA, the oldest professional women's organization in the country. "She seems to be more comfortable with herself, her position, her power."

"She understands it's still an all-boys club, and she's going to be a player," said Sen. Dennis DeConcini (D-AZ). "She already is."

MEANWHILE, EMILY'S LIST began to prepare for the 1988 election cycle, armed with the knowledge that it would set the stage for what EMILY's List was to become as an institution. I wanted help in evaluating our efforts, so I set out to get advice from allies and friends on what we had done and wanted to do in the future. Roger Craver, a direct-mail guru and a fellow progressive whom I had worked with at Common Cause, suggested we send a report to members of EMILY's List each election cycle so they would have a detailed sense of what we had accomplished in the past election:

how much money we spent on each program, what the staff assignments were — even how many pages of reports we sent to the Federal Election Commission. We had nothing to hide and were proud of how we were investing members' contributions. Our members were intelligent, sophisticated consumers, and it was important to show them there was a rational foundation for their investments. I firmly believed that this forthright approach would build trust and encourage our women donors to do more in the future.

To address the question of exactly how involved EMILY's List should be with the Democratic Party, I reached out to various political women and had long talks with Marie Bass, Joanne Howes, and Ranny Cooper, who, of all the women on our steering committee, were the most deeply involved in electoral politics. I was relieved when everyone seemed to agree that getting too close to the Democratic Party would be like stepping into quicksand.

On February 20, 1987, we also kicked off a new membership drive with a party at which Senator Mikulski came in to thank us. "We could not have won this wonderful campaign without EMILY's List," she said. "Your yeast helped me raise my dough. To know that you were there was a source of support for me." She added that she loved being in the Senate but she wanted some company. "I'm happy, but I'm lonely," she said.

Of course, getting her some company was precisely our mission, and we had already begun reviewing the upcoming 1988 Senate races to look for potential seats for women. But as we went through the list of states in alphabetical order, our spirits sagged. It wasn't until we got to Vermont that we saw a single possible opportunity. Sen. Robert Stafford was said to be retiring, which would leave an opening for Gov. Madeleine Kunin. We met with Governor Kunin several times. But when she decided to run again for governor, that meant that not a single Democratic woman would be running for the Senate that cycle.

Then, when we looked at the House, we came to a staggering realization: Democratic women had actually been losing ground.

Few people realized this, because between 1973 and 1986 the total number of women in the House had increased from sixteen to twenty-three. But that was only because the number of Republican women had soared from two to eleven. Most women's groups were bipartisan back then, so the darker truth about the Democrats had been buried — namely, that their numbers had actually *decreased* from fourteen to twelve. And if we couldn't reverse that trend, we would never have a large enough pool from which to elect women to the Senate — much less to groom women for cabinet posts or the White House.

Mind you, this was at a time when the Democratic Party had a ninety-one-seat majority in the House. All of which meant the party establishment was certainly not looking out for women candidates.

One of the first items on our agenda was getting our donors to renew. Conventional wisdom had it that we should expect only a 50 percent renewal rate — but we did much better. Out of 1,167 members, 752 signed up again in 1987 — 64 percent. As for recruiting new members, I embarked on a series of road trips — more than two dozen over the next two years — that eventually took me all across the country. First, in the spring of 1987, came Albuquerque, Atlanta, Austin, San Antonio, Los Angeles, Chicago, Phoenix, Boston, Des Moines, and Montgomery County, Maryland. Often, there were special guests in these destinations — usually a local Democratic woman politician of some note. Again and again, I made my pitch. "With such a small pool of women in the House, it is very difficult to move women into higher offices," I said. "We believe we can have an impact on the House by raising $50,000 for Democratic women to launch their campaigns. This large, early donation will have a phenomenal impact on the races, and will have a tremendous difference in increasing the number of women in the House."

In selecting which races to recommend, I pointed out, EMILY's List would use sophisticated research and a targeted approach to make sure that the money went only to viable candidates. And, of course, thanks to Senator Mikulski's historic victory, we had already proven how powerful EMILY's List could be.

As far as that selection process went, we had put together a comprehensive procedure for choosing viable House candidates to endorse in 1988. Because we were committed to delivering the most bang for the buck for our donors, that meant that, no matter how wonderful a candidate might be, EMILY's List was not going to support a woman who didn't have a chance. The first step was to make certain the opportunity was promising. Given that 98 percent of incumbents who ran were reelected, we were very skeptical about any candidate attempting to challenge a sitting member of Congress. At this early stage in EMILY's List, it was clear to us that open seats — seats for which an incumbent was not running — would be the easiest and most promising races to contest, especially those open seats in Democratic-leaning districts.

Next, we examined candidates in competitive districts. We sent each of them a comprehensive questionnaire asking them for their qualifications, specific campaign plans, a financial plan, and polling data about both their districts and their candidacies. After meeting them personally to review their races, we conferred with our political network across the country and in their home areas. Ultimately, the decisive factor again was viability — which candidates, if given early resources to build a campaign, had the ability to win — and more than a dozen were eliminated because they just didn't have a chance.

THEN, IN EARLY 1988, while I was on yet another West Coast swing to San Francisco, I met a remarkable forty-seven-year-old freshman congresswoman who was brand new to the House of Representatives but had been steeped in politics since birth. Nancy Patricia D'Alesandro had grown up in Baltimore's Little Italy as the daughter of Tommy D'Alesandro, a legendary Democratic congressman and, later, mayor of Baltimore. As a young girl, she was famously offered a toy elephant by a Republican poll worker, but

—true to her Democratic birthright—recoiled in horror and rejected it. While attending Trinity University in Washington (now known as Trinity Washington University), she met and later married Paul Pelosi, then a Georgetown student. In 1969, they moved to San Francisco, where Paul owned and operated Financial Leasing Services, a real estate and venture-capital investment and consulting firm.

Nancy Pelosi raised five children and worked her way up the volunteer ladder of Democratic politics in California, serving as a Democratic National Committee member, as party chair for Northern California, and then as party chair for the entire California Democratic Party. In 1986, as finance chair of the Democratic Senatorial Campaign Committee, she played a key role in helping the Democrats retake the Senate. Thanks to her considerable political acumen, she was a major player in Democratic politics in California for two decades, particularly in fund-raising.

In early 1987, Rep. Sala Burton (D-CA) was stricken with cancer and chose Pelosi as her designated successor. Burton died on February 1, and on April 7, Nancy Pelosi defeated San Francisco supervisor Harry Britt in the Democratic primary. She went on to win a special election in June. Her primary victory had been narrow, but, once elected, she had become ensconced in one of the safest Democratic districts in the entire country.

As the 1988 election season got under way, after she had been in Congress for about six months, Nancy generously hosted an EMILY's List fund-raising event at her house in San Francisco. When she introduced me, however, I discovered that she had an agenda of which I had been unaware. As the group assembled in her living room, Nancy said, "Ellen, I know EMILY's List will support good women running for the House, and I think you should announce your support for Anna Eshoo right now."

I was stunned.

Nancy was referring to her good friend and colleague who was running for Congress in Silicon Valley. Anna was a wonderful, warm woman who I believed would make a terrific member of Congress. I fully expected EMILY's List to endorse her. We had already begun our process of vetting her, and it seemed highly probable she would meet our criteria. But we had not finished that process. We had an important commitment to our members to assess viability before we asked members to support our candidates. With so few women in office, we could not afford to waste a dime of their money or ours on candidates who were unlikely to win. And as it was still very early in the election year, we had not yet determined who would make the cut. I wasn't going to be pushed into a corner.

When I delivered my spiel, I tried to respond to Nancy in the most politic way I could. "Anna's fantastic," I said. "She's terrific. And when we go through our process, I'm sure we will support her."

But that wasn't enough for Nancy. "I hear what you are saying, but here we all are, gathering to help EMILY's List," Nancy said. "Surely, you should endorse our good friend and neighbor, Anna Eshoo. I think you should tell us all right now that EMILY's List is going to be behind her."

The temperature in the room rose as I saw the steely look on Nancy's face. Many years later, when she became Speaker of the House, a reporter asked Nancy how she could possibly control the Democratic members of Congress. Nancy replied sweetly, "I just give them my mother-of-five stare." I was getting an early taste of exactly how forceful and persuasive that stare could be. I was anxious about how to handle this situation, but I was also impressed. Nancy Pelosi had an important political goal. She was being strategic and tough. She wouldn't let go. But we had a steering committee, the Founding Mothers of EMILY's List, who would decide whom we would support after we completed our evaluation. This wasn't solely my call, and it shouldn't be. So, I again praised Anna but said our official endorsement would have to wait. Then I quickly thanked everyone and ended the program.

★   ★   ★

ULTIMATELY, EMILY'S LIST did endorse Anna Eshoo — but only after dutifully observing its protocols. In addition, we decided that eight other candidates merited our endorsement for 1988. Our best opportunities, of course, were in open seats in Democratic districts. Patricia Madrid became a candidate in New Mexico's First District when Republican incumbent Manuel Lujan finally announced his retirement. We anticipated two open seats in Washington State, because two representatives were running for the Senate, and we joined with the Women's Campaign Fund to send political consultant Celinda Lake to Washington to recruit state senator Ruthe Ridder and state representative Jolene Unsoeld as candidates for those seats.

In Colorado, we backed Martha Ezzard, formerly a liberal Republican state senator, who was in the hunt as a socially progressive fiscal conservative running against Dan Schaefer, a Republican incumbent in a moderate district. In Syracuse, New York, Rosemary Pooler, who had almost beaten an incumbent in 1986, was running for an open seat. In Michigan, state senator Lana Pollack faced a six-term Republican incumbent who we believed was vulnerable. Similarly, in Louisiana, Faye Williams was challenging an incumbent we thought was weak.

ONE BIG DIFFERENCE between 1988 and the previous election was that we were now endorsing nine candidates, not just two. Our concern was that if we sent out mailings recommending all of them to our members and left it to the readers to decide, many people would find the process so complicated that they would bail out. We needed a more manageable way to allow people to participate.

"Because we wanted to distribute the money evenly, we just divided our membership list up alphabetically," said Betsy Crone. "One section would get the first two candidates we wanted to endorse, the second section would get another — and so on. Sometimes we recommended four candidates, and sometimes six. As the election progressed and candidates lost primaries, the list changed with each of the four or five mailings we did per cycle. Certain people looked more viable than others or needed more money, so we adjusted accordingly." It wasn't about candidates who lived close by or about whom members may have read in the newspaper. It was about who had a chance of winning so that we could increase our numbers in office. "When you elected Barbara Mikulski, you knew she was speaking not just for the citizens of Maryland, but for women all across the country," said Sherry Merfish, the Houston attorney who joined EMILY's List in 1986. "For those of us in Texas, where the landscape for women was brutal and utterly hopeless, where you didn't have a voice, I can't tell you how important it was to be able to elect a woman senator in another state who you knew was going to speak for all women. Those women became *our* senators. That feeling was just phenomenal."

As the 1988 electoral season got under way, EMILY's List–sponsored recruiting parties across the country became, for a number of women, a terrific entrée into a rewarding political career. In Lansing, Michigan, Debbie Stabenow, a thirty-seven-year-old state representative with a gorgeous mane of shocking red hair, hosted one of our parties after a Democratic woman friend approached her. Not long afterward, the big family room at Stabenow's house in Lansing was packed with enthusiastic women. "At that time, forking over $300 was no small thing," said Stabenow later. "It meant that you were really committed, that you wanted this to be taken seriously. It meant you were excited by the advancement of women in politics, and by seeing a wave of women who had started in local politics, on the school boards or county commissions, and who were moving on to state politics as I was at that time."

In March, our party in Houston, hosted by Sherry Merfish, was particularly memorable in that two special guests gave some indication of exactly how we were growing and why. Just thirty-two years old at the time, Mary Landrieu was the daughter of New Orleans mayor Moon Landrieu and, as such, a member of one of Louisiana's most prominent political families. First elected to the Louisiana House of Representatives at the ripe old age of twenty-four, she had just been elected state treasurer and had already set her sights on higher office.

Impressive as Mary may have been, it was the other guest at Sherry's Houston home who stole the show, holding forth with an extraordinary mixture of charisma, wit, and charm. As Texas state treasurer Ann Richards spoke, Sherry Merfish's two young daughters looked up at her, in rapt attention, leaving Sherry with the realization that there was indeed hope for the next generation of women — even in the political wilderness of Texas.

BORN DOROTHY ANN WILLIS, Ann Richards grew up in Waco, Texas, nearly two hundred miles northwest of Houston. The Willises were so poor that Ann's father, a deliveryman, compared the family to the Joads in the novel *The Grapes of Wrath*. A few hours after Ann was born, her mother got out of her birthing bed, wrung the neck of a chicken, and began preparing it for dinner. Ann frequently went without shoes. In later years, Ann made fun of the modest circumstances in which she grew up, complaining that "my mama and daddy did not have the foresight to arrange for a manger or a log cabin."

In the late fifties and early sixties, Ann and husband David Richards, a labor lawyer, were the heart of Dallas's minuscule liberal scene, picketing segregated restaurants like Piccadilly Cafeteria, and staging political parodies at which Ann sang songs she wrote and impersonated Lyndon Johnson in a vocal group called LBJ and the Sisty Uglers. A smart, polished, and attractive family with four children, the Richardses could have easily joined Dallas's country-club set — but they were rebels.

By the late sixties, the Richardses were so fed up with Dallas's buttoned-down conservatism that they fled to Austin, a refuge for the disaffected, from liberal intellectuals to redneck cowboys. As Craig Unger reported in the *New Republic*, one of their principal watering holes was Scholz Garten, the inspiration for the Dearly

Beloved Beer and Garden Party of William Brammer's legendary political novel *The Gay Place*. "This was where you tapped into the zeitgeist of Austin," said David Richards, Ann's husband. Guests at the Richardses' house ranged from Dennis Hopper to *New York Times* editor Abe Rosenthal. "Ann and David were the Dick and Nicole Diver [the dazzling couple at the center of F. Scott Fitzgerald's *Tender Is the Night*] of the scene, gay and glamorous — the kind of people who made things fun when they walked into the room," said Ann's friend the author and journalist Molly Ivins.

Colorful though it was, theirs was a fringe crowd that almost seemed to delight in losing its political battles and being ironic about it. Power was anathema. Or so it seemed until Ann began to argue that it was time to get out of the back room and take part in politics directly. She was one of the founders of the National Women's Political Caucus in Texas, and traveled the state organizing women and teaching young girls about the contributions of women throughout history. In 1972, she worked on Sarah Weddington's successful run for the Texas state legislature, just after Weddington, an Austin lawyer, argued the historic case before the United States Supreme Court known as *Roe v. Wade*.

In 1975, David Richards was approached to run for commissioner of Travis County, where Austin is located, but David, an effective but uncompromising idealist, stepped aside in favor of Ann and her consensus-building pragmatism. Ann courted local power brokers such as University of Texas Board of Regents chairman Frank Erwin, Lt. Gov. Bill Hobby, and Sen. Lloyd Bentsen. When she finally ran for office herself, she was able to use these relationships to win support from the business community without alienating her liberal base. "She understood that women have to establish themselves with powerful male insiders as credible candidates or they will have a tough if not impossible time of it," said Jane Hickie, an Austin-based attorney who was Ann's close friend for more than thirty years and became her right-hand adviser.

Ann won the commissioner race handily. It was an unusual job for a woman, "a truck-driving, front-end-loader" kind of job, she called it, which involved managing road crews and the like. But Ann did so well that it quickly became clear she was destined for greater things.

NOT THAT SHE WAS a flawless candidate. In 1980, Ann's family confronted her about her drinking, and after a stint at Hazelden, the Minnesota substance-abuse treatment center, Ann returned to Austin, a recovering alcoholic. She and David soon divorced. "Back then, a divorcée who was in rehab, well, that would lead you to say she was not the ideal candidate," said Jennifer Treat, who worked with Ann for many years and was her Washington-based fundraiser.

In 1982, the Texas Democratic Party finally patched up an age-old split between its conservative and liberal wings. Democrats ranging from moderates like Mark White, as governor, to maverick populist Jim Hightower, as agriculture commissioner, swept into office. Among them, Ann was elected state treasurer and became the first woman elected statewide in fifty years — winning more votes than any candidate on the entire ballot.

In her first term, Ann restructured the state's archaic revenue system and helped Texas bring in an extra $2 billion in nontax income. "Ann took over an agency that was a moribund, rust-infested pit, and made it work," said Molly Ivins. She was so successful that she won her second term unopposed.

Having grown up surrounded by hard-drinking good old boys, Ann could talk pure Texas, and, as one friend put it, she told stories "that would make you blush and run." Over time, she cobbled together a persona that was a mesmerizing amalgam of feminine Texas archetypes, at once funny and charming, folksy and tough. What made her extraordinary was that because she was genuinely steeped in the mythology of Texas, she was masterful at using her acid wit to expose the sexism, racism, and hypocrisy that often lurked beneath.

When Ann appeared before women's political groups, she was known to don a rubber pig's snout and transform herself into Harry Porko, the archetypical Texas sexist. Similarly, as Molly Ivins reported, Ann and she once attended a high-powered political event at Scholz Garten in Austin with Bob Bullock, who was then the state comptroller, and a black man named Charlie Miles, who headed Bullock's personnel department. Before long, a "sorry, no-account sumbitch" judge, as Ivins described him, approached Bullock and clapped him on the back. "Bob, my boy, how are yew?" said the judge, who had a reputation for being something of a racist.

"Judge, I want you to meet my friends," said Bullock, introducing him first to Molly.

Next, Bullock introduced Miles. "This is Charlie," he said, "who heads my personnel department."

Charlie Miles extended his hand. For a moment, as Molly put it, the judge just looked at the hand "as if he had just stepped into a fresh cow pie."

Finally, after grudgingly shaking hands with Miles, the judge hurriedly turned his attention to Ann, who happened to look stunning that day. "And who is this lovely lady?" he asked.

Ann smiled and didn't miss a beat. "I'm *Mrs.* Miles," she said.

As she climbed the political ladder, Ann incorporated her uniquely Texas sense of irony into her speaking engagements. She honed her skills at such Texas venues as the Madisonville Cattleman's Association, the McAllen Mexican-American Democrats convention, and the Nacogdoches Chamber of Commerce. She was adept at pointing out the foibles of her own party — especially when it came to the absence of women and minorities. "Are there any other white men who haven't spoken yet?" Ann asked after she took the podium at one Democratic fund-raiser. "God knows we wouldn't want to leave any of you out."

"She was unbelievably intelligent," said Jennifer Treat, "not so much in a cerebral way, but as in having an innate instinct, an ability to read the audience and adapt and recalibrate instantly." All of which made her, at fifty-four years old, a rising star on the cusp of gaining national attention.

AS EMILY'S LIST'S ROLE in the electoral process grew, not surprisingly, we began to attract more attention. In June, the *Washington Post Magazine* ran a four-thousand-word piece on us, "Women Can Be Power Brokers, Too; How Ellen Malcolm Learned to Influence Elections — and Love It," by Amanda Spake. We had gotten press coverage before, of course, but this was the first lengthy report in a major outlet about what we had accomplished. Now, the *Washington Post* said, we had "clout." EMILY's List was the Democratic Party's "leading queenmaker."

Meanwhile, with Vice President George H. W. Bush, a Texan, running for what was in effect the third term of a popular Ronald Reagan presidency, the leading Democratic candidate, Gary Hart, had withdrawn after reports of an extramarital affair surfaced in the *Miami Herald.* After a host of potentially strong candidates — New York governor Mario Cuomo, Arkansas governor Bill Clinton, New Jersey senator Bill Bradley, Massachusetts senator Teddy Kennedy, and Texas senator Lloyd Bentsen — chose not to run, Massachusetts governor Michael Dukakis moved to the head of the pack and swept the primaries.

To balance the fact that a liberal New Englander was at the top of the ticket, and perhaps in hopes of rekindling the Massachusetts-Texas axis embodied by the JFK-LBJ ticket, Dukakis chose Lloyd Bentsen as his running mate. In case that wasn't enough to help win Texas's twenty-nine electoral votes, the Democrats went a step further.

In June, as Ann Richards and her friend Jane Hickie were passing through Houston's Hobby International Airport, Hickie's shoe box–sized cell phone suddenly rang. Paul Kirk, chairman of the Democratic National Committee, was on the line, asking whether Ann would give the keynote speech for the upcoming convention. Potentially, Ann could boost the Democrats' chances not just in

Texas but in the entire South and with women all across the country as well.

A star debater in high school and college, Ann had enjoyed hundreds of speaking engagements as a politician, but this was her first time on the national stage. Her team immediately started calling everyone who might help. Ann consulted with more than a dozen speechwriters, politicians, and journalists, including comedian Lily Tomlin, New York governor Mario Cuomo, JFK speechwriter Ted Sorensen, LBJ aides Harry McPherson and George Christian, and Barbara Jordan, the civil rights leader and congresswoman who had delivered a memorable keynote speech twelve years earlier.

At the behest of Bob Strauss, former chairman of the Democratic Party, Ann hired speechwriter John Sherman, who immediately went to work. And Lily Tomlin came up with what may have been the single most important piece of advice: lighting is crucial. You have to make sure the house lights are turned down. You don't want the TV to focus on delegates wearing funny hats. You want the focus on Ann. So, Ann's team spoke to the networks, and they agreed to dim the house lights.

Meanwhile, as Ann busily prepared for the speech of her life, the entire country speculated about what she would say. Political consultant Harrison Hickman thought the speech should be high-minded. Others thought Ann should be allowed to be Ann. Suddenly, there were too many cooks in the kitchen. In the *Washington Post,* Lois Romano noted that Ann's friends were anxious that she might forgo "her trademark humor and choose instead to use this star-making opportunity to be — Heaven forbid — serious." What they wanted instead was the Ann Richards they knew and loved, the "brassy quipster [with] . . . the kind of wicked wit that starts out fast and peaks at such a lightning speed, her audiences are reeling for days."

Then, just before Ann and her entourage left to go to the convention in Atlanta, as John Sherman finished revising his final draft, the computer "ate" the speech. It was gone forever, nowhere to be seen.

When she arrived in Atlanta, Ann was stunned by the glare of the media. "We were completely unprepared for the avalanche of attention before, during, and after the event," said Jane Hickie.

"You feel like they've made a mistake," Ann said. "That they really don't understand that you're not that important. It was my first true moment in the eye of the media storm, and it took some getting used to."

Sherman and his team worked around the clock from old notes and early drafts to put the speech back together again. But Ann wasn't satisfied. She knew she had to walk the line between the gravity called for on the occasion and the breezy good-old-girl style she was famous for. "I wanted to speak so that my Mama understood what I was talking about," she explained in her autobiography. "Right from the beginning of the speech I wanted to make it clear that 'We're about to have some fun.' I wanted an overall feeling that made people know that politics does not have to be all gloom and doom and lofty rhetoric, that it is really personal, and that it's fun."

MEANWHILE, I MADE the trek to Atlanta to attend my first national political convention. EMILY's List was still brand new, of course, so my goal was to make sure everyone in the Democratic Party found out about us. I was a woman on a mission, fixated on telling as many people about EMILY's List as possible. To that end, we had hundreds of buttons made saying I'M ON EMILY'S LIST, an intentionally vague message designed to get people to ask about us.

At the time, I had no idea how conventions work. All the hotels were already booked before a group of us made plans, so we ended up renting a house on the outskirts of town. Old hands at politics know how to hustle tickets for all the important events, but I was clueless. So, on the evening of Monday, July 18, I stayed at our rental and, like tens of millions of other Americans, watched Ann Richards on TV.

In terms of my own personal progress as a public speaker, the convention was a great opportunity to study the techniques of Ann and civil rights leader Jesse Jackson. Jesse had a way of taking a word or a phrase — "common ground!" — and repeating it throughout the speech until it built to a crescendo. It was extraordinarily effective.

I also realized that Ann's distinctive style and sense of timing made her speeches far more effective when she delivered them than when the same words appeared on paper. She used her Texas drawl to draw out her best lines slowly, so that you were on the edge of your seat, waiting for the punch line. You knew something good was coming, but you wondered what it was going to be. She had mastered the art of creating suspense.

On the day of the speech, Ann dropped by the CBS News booth in the convention hall, where Walter Cronkite, the legendary newsman, was anchoring the ceremonies. Ann had known him for years. "Walter," she said, "I want you to be prepared for what kind of speech you're going to hear from me tonight."

Cronkite looked at her quizzically.

"I'm going to talk Texas," she said.

"Well," he replied, "that's great."

Ann's visit with Cronkite was not a mere social call. When Lyndon Johnson had been president, his accent had been reviled and caricatured as the utterings of an uneducated, backwoods redneck, and Ann wanted to make certain she wasn't tarred with any such associations. Cronkite had attended high school in Houston and had gone to the University of Texas, so he knew better. "A Texas accent is not something Americans appreciate," said Jane Hickie. "She wanted Cronkite to prepare the press for a different kind of speech."

And so, on the evening of July 18, wearing a double strand of pearls around her neck, a light-blue dress, and her perfectly coiffed crown of white hair, Ann prepared to take the podium at the Omni Coliseum in Atlanta. But minutes before she was scheduled to take the stage, something happened that threatened to undo everything.

A representative of the networks told Ann that they had decided to *not* turn down the house lights. They were reneging on a promise that was absolutely crucial to the entire production.

To a seasoned performer like Lily Tomlin, turning down the house lights was standard stagecraft. At a political convention, where everyone comes to see and be seen, where many people are not paying attention to the speaker, and where the TV cameras are on twenty thousand people wearing silly hats, waving signs, and hoisting balloons, it is even more important. There is always someone yawning, sleeping, or picking his nose, and with the lights up, bored cameramen can easily undermine even the most eloquent oratory.

But Ann knew exactly how to handle it. "If you don't turn down the lights," she said, "I won't talk."

It was as simple as that. Ann made it crystal clear that there was nothing more to say. There would be no negotiating. The networks would do as she said or face thirty-five minutes of dead air in prime time on national TV. "She was so poised, so composed," said Jane Hickie, who was present.

A few minutes later, the lights dimmed. Ann went out onstage behind the podium. She smiled at the crowd, and her distinctive Texas drawl filled the arena. She spoke slowly.

"I am delighted to be here with you this evening," Ann said. There were pauses not just between sentences, but within them. "Because after listening to George Bush all these years, I figured you needed to know what a real Texas accent sounds like."

There was laughter, and she waited until it subsided.

"Twelve years ago, Barbara Jordan, another Texas woman, made the keynote address to this convention," she continued.

Pause.

"And two women in 160 years is about par for the course."

There was more laughter.

"But, if you give us a chance, we can perform."

Pause.

"After all, Ginger Rogers did everything that Fred Astaire did."

Pause.

"She just did it backwards and in high heels."

The entire crowd roared—with one notable exception. By and large the party elders, particularly Texans such as Speaker of the House Jim Wright and Sen. Lloyd Bentsen, had celebrated Ann's selection as keynote speaker. But one man was not happy about it. Jim Mattox, the Democratic attorney general of Texas, planned to run for governor in 1990, and he knew nationwide exposure would give Ann a significant leg up if she ran as well. Now he was watching his worst nightmare unfold in real time, the meteoric rise of a woman who was about to become his chief rival.

A cameraman from Texas realized what was going on and had trained the camera on Mattox. "He was in despair," said Jane Hickie.

Oblivious to Mattox, Ann continued in an intimate tone, speaking to twenty thousand people in the arena—and tens of millions across the country—as if she were talking to them in their living room, as a friend, one-on-one. Over the next thirty-five minutes, she proceeded to touch on the all the vital issues the Democratic Party had fought for over the years—health care, Social Security, the forgotten middle class, the family farm, and labor. As the *Los Angeles Times* put it, she summoned up deeply personal memories "of her Depression-era small-town Texas childhood, where kids sat on quilts under the stars at night and listened to grown-ups, and her mom put Clorox in the well when a frog fell in." She made millions of people identify with her as an ordinary human being. Her timing was exquisite.

She talked about Republicans as the divide-and-conquer party. Then, she went after George H. W. Bush—mercilessly. For eight years, she said, Bush had pretended he could not hear the questions of the press over the noise of the helicopter. On the rare occasions he bothered to answer questions, she said, he gave nonresponsive answers we wouldn't have accepted from our own children. She assailed Bush's sense of entitlement, his political opportunism, his newfound sense of caring, now that he was finally looking for a job he couldn't get appointed to. Then came the kicker.

"Poor George," she said.

She paused for six seconds, while the audience laughed.

"He can't help it."

There was more laughter, and another pause, this one a few seconds longer. This was theater, and Ann was giving an award-winning performance. She knew exactly how to handle it, when to turn it on, and when to turn it off. I had often played bridge with her, and, in a casual setting like that, I barely noticed much of a Texas accent. But this was a moment when she knew she should pour every ounce of Texas she had into the next sentence.

"He was born with a silver foot in his mouth."

As *The New Yorker* put it, with that line, "the roar of the crowd came in like breakers on the shore. You could hear the wave rise as she unreeled her line, and then it crashed on her and drenched her." The cheering and laughter continued for a full forty seconds, then receded.

"I'm a grandmother now. I have one nearly perfect granddaughter named Lily," she continued. "And when I hold that grandbaby, I feel the continuity of life that unites us." With that, the crowd stood up again, and the cheering seemed to go on forever.

When she went backstage, Ann, who was troubled by the unfounded concern that she had not made eye contact with the camera, immediately ran into broadcaster Diane Sawyer, then with CBS News. "Did I do okay?" she asked.

Ann had done far more than okay. In one speech, she had dazzled America with two lines—one about Fred and Ginger, the other about George's silver foot—that would be repeated again and again for decades.

A star had been born. As the *Houston Chronicle* wrote, she had "metamorphosed from third-rung public official to a sought-after national figure, the toast of the Democratic Party."

CBS anchor Dan Rather put it succinctly: "I heard people at the convention say, 'Whew. We've got the wrong person running for president.'"

Nine

HOW TO BEAT BUBBA

THE NEXT MORNING, Tuesday, July 19, EMILY's List joined with the National Women's Political Caucus and the Women's Campaign Fund in a presentation featuring Ann and vice presidential nominee Lloyd Bentsen. I introduced Ann, to thunderous applause. Bentsen was delayed, which meant that Ann had to vamp on and on — which she did incredibly well. It was clear she was being transformed from a mere state official into a significant national figure. International, even. The BBC wanted to fly her over to London. Xinhua, the Chinese news agency, was lauding her. State treasurers simply did not get that kind of attention.

But a few days later, Ann returned to Austin and received a reception that was stunningly at odds with the worldwide adulation she was receiving. Not long after her arrival, she got a call from George Christian, one of the more powerful men in the Texas Democratic Party. A former journalist, Christian had had served as press secretary for two Texas governors before joining Lyndon Johnson's staff in the White House. When Bill Moyers resigned as LBJ's press secretary, Christian took that thankless but highly visible job for three turbulent years in the White House. He subsequently returned to Texas as a major power broker in the state Democratic Party.

Now, he insisted Ann come to his office. Immediately. It was important.

Ann and Jane Hickie soon arrived at Christian's office, meeting in a small conference room in a downtown Austin bank building, but neither of them knew the purpose of the meeting. In addition to Christian, there were George Shipley, a legendary political operative who was known as Doctor Dirt for his opposition research on Republicans; Jack Martin, the longtime campaign manager and aide to Sen. Lloyd Bentsen; former journalist-turned-political-strategist John Rogers; and three or four others.

Between them, Christian, Shipley, and company represented the Texas Democratic Party establishment. They were the men behind every major Democrat in Texas politics, from Gov. John Connally and Lyndon Johnson to Gov. Mark White and Lloyd Bentsen. They were the Texas equivalent of the men in Missouri who had left Harriett Woods high and dry, the counterparts of the Maryland Democratic establishment that viewed Mike Barnes as "a golden boy" in the House who was more senatorial than Barbara Mikulski. And now, at a time when Ann Richards was being lionized by millions of people all the way from her hometown of Waco, Texas, to Beijing, when even president-elect George H. W. Bush, the subject of her barbs, had been gracious and good humored enough to send a small silver foot as a gift, these men were calling her on the carpet. To put it mildly, they were not one bit happy.

Ann didn't have a clue about what was going on. "It was like, 'Whoa,'" said Jane. "Ann was sort of shocked. Everyone was celebrating her and here were these powerful men who were supposed to be behind her, but they were the first to tell her she had done something wrong."

What could Ann have possibly done wrong? These were her political allies. It wasn't as if she had tried to leave them out of the loop. Just after she was asked to speak before the convention, Ann phoned Christian for advice, but, Jane said, he had been somewhat dismissive.

Before long, the purpose of the meeting became crystal clear. Ann had become a much bigger star than anyone had anticipated.

More to the point, her sin had been that she had gotten there on her own — *without* their help. And that was a state of affairs that would not be allowed to continue.

"It was a real dressing down," said Jane. "It frightened her. When Ann left that office, she was ashen."

The bottom line: if Ann was to run for governor, she had to make room on her team for these men and their surrogates. As a result, she hired Shipley to do "oppo" research. Jack Martin became her campaign treasurer and, perhaps more importantly, served as a reassuring high-profile presence in the media. Finally, as campaign manager Ann hired Glenn Smith, a former journalist who had worked for Lt. Gov. Bill Hobby and, subsequently, Sen. Lloyd Bentsen. He had not been present at the meeting, but he stood in good stead with the powers that be. The old boys were now wired into Ann's campaign. They had reasserted their primacy.

"Ann absolutely had to accommodate them," said Jane. "Their guys had to be in charge."

ANN'S ORATORICAL TRIUMPH NOTWITHSTANDING, in 1988, as in 1984, the national Democratic ticket once again went down to a resounding defeat, with George H. W. Bush and Dan Quayle beating Michael Dukakis and Lloyd Bentsen by 8 points in the popular vote and 426 to 111 in the Electoral College. In spite of the setbacks at the top of the ticket, EMILY's List managed to make history once again.

Not that it was easy. One of our victories came in the Third Congressional District in Washington State with a woman who was anything but a conventional politician. The widow of mountaineer Willi Unsoeld, Jolene Unsoeld had spent two years as director of the English Language Institute in Kathmandu, Nepal, before returning to the United States and being elected as a state representative. There, she became the conscience of the state legislature by lobbying for open government, campaign-finance reform, and environmental issues.

Jolene had little political experience, but we walked with her every step of the way. In early 1987, we had identified her as a potential candidate to replace Democratic incumbent Don Bonker once we learned he had decided to run for the Senate. We hired pollster Celinda Lake to give Jolene advice. We helped her raise money for radio ads for her primary battle, and when she won that, we helped fund TV ads for her for the general election.

Nita Lowey was another story. Initially, we had been wary of endorsing Nita to run in New York's Westchester County, though not because of any reservations about her personally. As assistant secretary of state for the state of New York, she had no electoral experience and was running against Joseph DioGuardi, the well-funded Republican incumbent. In addition, she had a competitive primary against several well-known and well-funded Democrats. All of which made her such a long shot that at first we decided she simply did not pass the viability test.

We knew Nita was warm, personable, and an excellent campaigner. But it turned out that she was also affluent enough to help finance her own campaign with a $250,000 loan, and as a result she came from behind to win her primary. We were flabbergasted, and immediately endorsed her. Because it was a September primary, we did not have time for a candidate mailing, so we profiled Nita in our newsletter, asked members to send her checks, and contributed $5,000 directly to her campaign. But she was still given little chance of beating DioGuardi in November.

Then, about three weeks before the election, Gannett Westchester Newspapers revealed that twenty-nine employees of an automotive company owned by a DioGuardi supporter were offered checks of $2,000 each and then were asked to contribute the same amount to DioGuardi campaign committees — in violation of federal election laws. With DioGuardi on the defensive and surrounded by scandal, the *New York Times* endorsed Nita.

Come election night, Jolene Unsoeld and Nita Lowey were both neck and neck with their rivals. Ultimately, Jolene won in a squeaker

— by exactly 657 votes. That called for a recount and, to help out, we sent an organizer so she wouldn't be overwhelmed. Similarly, Nita eked by with a 2,000-vote margin.

EMILY had made history once again. This time around, our members had contributed a record $600,000, making us the biggest donor to women candidates in the entire country. Most importantly, we had reversed the historic decline of Democratic women in Congress. Fourteen Democratic women were now taking their seats in Congress — the highest number ever.

AT ABOUT THE SAME TIME, just after the election, a remarkable woman showed up at the EMILY's List offices. With her hair nicely coiffed and her nails newly done, she was so stylish she looked like she had just walked out of the pages of *Glamour* magazine. She certainly didn't look like the rest of us working in our shabby offices at 2000 P Street.

Judi Kanter was her name. She had volunteered for the Dukakis presidential campaign in San Francisco and, after being told by a twenty-two-year-old intern there that there was no job for her, said, "I'll be there to start work at nine tomorrow morning." Kanter ended up being in charge of three thousand volunteers for Dukakis. She loved it, and having just heard about EMILY's List, she wanted to help. After chatting with her for a while, I had to ask what got her so excited about us. Without missing a beat, Judi uttered two words.

"No bylaws!" she said. I laughed.

Judi was excited, I realized, because she saw that EMILY's List was not going to get bogged down in process. All our energies were going into achieving our goals. Yes, we had passed bylaws so we could incorporate, but that was about the only time anyone ever looked at them. I was determined that we would not be like organizations that spent most of their energy on internal politics. We were all about making a difference. Judi got it.

To convince me to hire her, she said she would organize an event or two in San Francisco, which, at that time, was a difficult place to raise political money. I had spoken there once but had signed up only a handful of members. I thought we might as well give Judi a try.

In early 1989, we hired our first executive director, Rosa DeLauro, the former campaign manager and chief of staff to Sen. Chris Dodd (D-CT). Bringing Rosa on board was crucial, because a gigantic opportunity was staring us in the face. The upcoming electoral season of 1990 marked one of those decennial years after which congressional seats are reapportioned following the census. Reapportionment then leads to redistricting, a procedure in which each state redraws congressional district lines.

Those shifting political districts, in turn, create an unusual number of open seats in the election that follows the census. In most election cycles, the House of Representatives has only thirty or forty open seats when an incumbent is not running for reelection. But in the "2" years — 1992, 2002, 2012, and so on — there may be as many as seventy-five to one hundred open seats. Not every state would lose or gain seats because of redistricting, but the redrawing of political boundaries can make or break dozens of politicians. So, for us, open seats meant political opportunities for women. In fact, since 1970, the *only* times the number of Democratic women in the House had increased were in "2" elections — 1972 and 1982. I had hired Rosa precisely so that we would be well positioned to take advantage of them.

To start with, Rosa headed up a project to identify as many as eighty districts that would be opened up by redistricting. We had walked every step of the way with Jolene Unsoeld, and we intended to do exactly the same with far, far more candidates once those district lines were redrawn. We also set a goal to raise $1 million for candidates in 1990 — two-thirds again as much as we had raised in 1988.

With that in mind and with Judi Kanter leading the way, we started a Bay Area Brigade in January. "I knew what I was doing," said Judi. "I pulled together four parties, and Ellen could see it."

To help kick things off, Ann Richards served as our guest of honor at a large donor dinner in San Francisco, where her acid comments about Secretary of Defense John Tower — "To know him is to detest him" — caught the ear of *San Francisco Chronicle* columnist Herb Caen. That got our phones ringing off the hook. Then, we boldly sauntered forth to other venues. All in all, the Bay Area effort increased our rolls by more than 7 percent, bringing in 150 new members. It was an auspicious beginning to what was, for me, a twenty-city membership drive all across the country.

But raising $1 million required more than simply expanding the EMILY's List membership base. It meant that it was time to upgrade — that is, to ask for more money from some of our more affluent members. We had been using the $100 figure as an effective marketing device, but every once in a while, we would look at our membership list and see that one of our members was on the board of a symphony or some charitable organization. Some of our members could obviously give more.

To that end, we launched the Majority Council for major donors, members who contributed $1,000 and up. I was on the board of the Human Rights Campaign Fund (now known as the Human Rights Campaign), and I happened to find out that a fellow member there, Edie Cofrin, was also a member of EMILY's List. It also turned out that Edie and her two sisters, Gladys and Mary Ann, were heiresses to the Fort Howard Corporation, at the time one of the largest manufacturers of paper products in the United States.

Our new executive director, Rosa DeLauro, wanted to hire more staff, so EMILY's List needed more money — and now was the time to ask for more. Up until that point, Edie had been a typical member who contributed $100 in checks each year. But I finally summoned up my courage to move forward. We met, we hit it off, and we set a phone date to talk in more detail about her becoming a bigger contributor.

On the phone, I cut to the chase. "Edie," I said, "we want to develop a political program so we can do more for the candidates than just raise money for them. I was hoping you and your two sisters, Gladys and Mary Ann, would contribute $50,000 to launch the program. Would you do that?" I had never asked for that kind of money and was nervous. But I was confident we could put the money to good use, so I was willing to forge ahead.

"Do you mean $50,000 combined," Edie asked, "or for all three of us to give $50,000 each?"

I hadn't really considered asking *each* sister to give that amount. But I knew an opening when I saw one. I was in uncharted waters. But there's an old saw in fund-raising that you get only what you ask for.

I paused for a moment. "Well, I was hoping each of you would give $50,000 . . . ," I said.

"I'll talk to them and get back to you," said Edie.

Two days later, Edie called me. The answer was "yes." I had turned a $100 donor into a $150,000 contribution. I was overwhelmed by the Cofrins' generosity. As soon as the call was over, I ran into to Rosa's office to give her the good news.

AS ROSA AND I AGREED, money was crucial, but it alone was not the answer. We had to do hard-nosed political planning for the next several election cycles. That meant working with national and state activists to find candidates for the 1990, 1992, and 1994 elections and to help them plan their campaigns.

To that end, in March 1989, we held a three-day debriefing conference with the women candidates of 1988, their campaign managers, and their staff to find out what we had done right and what we had done wrong. What campaign dynamics were unique to women candidates? Which techniques worked, and which didn't? How do women fare in the media? Exactly what role should EMILY's List be playing?

And there was one other development. One of the few bright spots now that the Bush-Quayle administration was in power was that since Dan Quayle had become vice president, Rep. Dan Coats (R-IN) had run for Quayle's Senate seat, meaning that Coats's congressional seat in Indiana had opened up.

In March 1989, Democrat Jill Long ran for the seat. She was considered a long shot, but many of us at EMILY's List were determined that Jill would win, and Rosa DeLauro feverishly worked her contacts in the Democratic coalition. With the help of EMILY's List, Jill squeaked by her Republican opponent by 1,775 votes. We had now increased the number of Democratic women in Congress by three. All three had won by fewer than 2,000 votes — but they had won, and we had increased Democratic women in the House by 25 percent in three years. Needless to say, we were excited by our future prospects.

THAT SUMMER, A RULING by the United States Supreme Court energized our members and helped our candidates. On July 3, 1989, the Court upheld a Missouri law that restricted the use of state resources in performing, assisting, or counseling on abortions. The ruling, known as *Webster v. Reproductive Health Services,* was a historic breakthrough for conservatives in that it had previously been thought that such legislation was forbidden under *Roe v. Wade.* Now, it wasn't clear exactly what restrictions could be imposed, but the door had been opened for many, many legal battles to come. This was real erosion, and it meant that pro-choice forces had to start organizing.

In response, we commissioned a poll to see, in the wake of *Webster,* whether pro-choice positions would help or hurt Democratic candidates come election time. The data confirmed that being pro-choice was a powerful asset for Democratic candidates, and the strongest candidate to run against anti-choice Republicans would be a pro-choice Democratic woman. Until this point, abortion had been a topic that pro-choice politicians had downplayed or ignored. We helped give them confidence to speak out on the issue. Unfortunately, *Webster* also gave great momentum to conservatives who were determined to put so many constraints on *Roe v. Wade* as to render it meaningless. The abortion wars had begun.

MEANWHILE, ANN RICHARDS announced her candidacy for governor of Texas. At the time, her approval ratings were astoundingly high — 73 percent positive to 7 percent negative. But even so, winning the Democratic nomination was by no means a foregone conclusion. As Mary Beth Rogers, Ann's chief of staff for two years, put it, "She was an underfunded, divorced single mother, a recovering alcoholic, a liberal Democrat, a civil rights activist, and a pro-choice feminist in a conservative, increasingly Republican-leaning state." Her Democratic rivals included former governor Mark White and Jim Mattox, who, as Texas attorney general, had so much power, the *Fort Worth Star-Telegram* noted, that "politically active establishment types dare not slight him." The result was that, by mid-1989, Mattox had put together a war chest of $3.7 million — nearly four times what Ann had raised.

As Democratic primaries go, the tenor of this campaign was particularly brutal. That was largely because of Mattox, the "junkyard dog" of Texas politics, who, as Molly Ivins put it, was so mean "he wouldn't spit in your ear if your brain was on fire." Mattox's political sensibilities were such that he went on TV and announced he liked the death penalty so much that, as attorney general, he used to go to the state penitentiary and watch guys get electrocuted just for fun.

Not to be outdone, Mark White asserted that only governors, not attorneys general, deserved credit for executions. To convince voters that he prized capital punishment even more than Mattox, he produced TV ads in which he strolled leisurely among posters of those who had been put to death. All of which left Ann with little to say except that "the death penalty is necessary and is the law of the land." Coming out against capital punishment would be political suicide.

For Mattox, the battle against Ann was personal. Just a few days into the campaign, he ran into Glenn Smith at the statehouse in Austin. Smith, who had just signed on as Ann's campaign manager, reached out to shake hands. "Don't go trying to be friendly with me," said Mattox. "You don't understand: This is personal war. You are now a lifelong mortal enemy of mine! And go tell everyone else thinking of working for Ann that I said it."

Suddenly, Ann's life had changed. Everything that had once been private was now available for public scrutiny. She had already addressed her alcoholism, and her recovery, in her 1989 memoir. But that wasn't enough. Now it was open season. Again and again, she was hounded by rumors that she had used cocaine.

Mattox poured fuel on the fire. "There is sufficient information to cause me to believe that she needs to come forward and discuss these issues in a public forum," he told a news conference. "She doesn't want to answer simple 'yes or no' questions about cocaine, marijuana, and hallucinogens. She must answer what she used, how much, for how long, and who supplied them." Because Ann declined to answer the charges, the issue simply would not go away. Mattox was relentless. It continued for months with an unending stream of negative ads on TV.

In addition to the fact that EMILY's List members were contributing more than $200,000 to Ann's primary race — more than any single contributor — we helped Ann develop a national fund-raising machine with more than twenty thousand contributors. "EMILY's List was absolutely critical," said Jane Hickie. "They came up with money we had to have, because Mattox had trial-lawyer money, and Mark White had money from the many friends he had made as governor. The line we used was, whatever you pay for shoes, write a check to Ann for the same amount. And because EMILY's List could do that all over the country, it was effective."

Initially, for all his money, Mattox made little headway in the polls. An early 1990 poll gave Ann 35 percent, Mark White 29 percent, and Mattox only 10 percent. But that only increased the attacks.

In March, in a three-way debate among the Democratic candidates, White and Mattox repeatedly badgered Ann with the question of illegal drug use.

"I have revealed more about my personal life, including my alcoholism and my recovery—for ten years—than any person who has ever run for governor before," she replied. ". . . By continuing to raise these questions, I think we are sending a very sad message to a lot of people that if they seek treatment they will forever bear the stigma of their addiction."

But Mattox kept hammering away. The press insisted that Ann answer the charges. Her negatives skyrocketed. The breezy, down-home, folksy good old girl was gone. "She was afraid," said Jan Jarboe, senior editor at the *Texas Monthly*. "She was like a punch-drunk fighter. Everywhere she went, she knew she was going to get hit with the question. So she stopped going places, and, for a while, was nowhere to be found."

She sank in the polls. On March 3, one poll showed Ann in third place, with the primary just ten days away. The next day, the *New York Times* reported that she was flailing about with her back against the wall. Many thought it was over.

Then, on Monday, March 5, at a press conference with two hundred reporters, cameramen, and supporters, Ann let loose with everything she had. "I said just what I intended to say about my alcoholism during the debate and afterward," she announced. "I haven't had a mood-altering chemical for ten years."

Then, she let Mattox and White have it, with a barrage of questions about Mattox's questionable contributions and White's wealth. Suddenly, a wave of aggressive, negative commercials from Ann attacking her rivals filled the air. Just two days before the primary, she inched ahead in a three-way logjam with 28 percent of the vote. Mattox and White were close behind, at 25 percent and 24 percent respectively.

Finally, when the votes were counted on March 13, Ann won 39 percent of the vote, Mattox 37 percent, and White 19 percent. Ann had held on, but she still needed to win a runoff to nail down the Democratic nomination.

MARK WHITE WAS NO LONGER in the race, but that didn't make things much easier. He denounced Ann and even compared her to Heinrich Himmler, one of the persons most directly responsible for the Holocaust.

By this time, it seemed as if the electorate was beginning to feel the same way. On election night, April 11, we knew things were looking good as returns rolled in and Ann won county after county in rural West Texas, a stronghold of machismo. Even in Lubbock, a conservative town in the Texas Panhandle, Ann was winning 65 percent of the vote. When all the votes were counted, Ann had beaten Mattox soundly, 57 to 43 percent.

The only problem was that the primary had been so brutal that Ann emerged battered and broke. She had been so busy fending off allegations of drug use, she had gotten no positive message across. Her negatives had soared from a mere 4 percent to 50 percent, and she was starting off the race for the general election fully 26 points behind Clayton "Claytie" Williams, the Republican nominee. No one laid a glove on Williams in the Republican primary, and he peppered the airwaves with nostalgic paeans to the frontier. If Ann was expecting any respite during the general election, she didn't get it.

What happened next, as CBS newsman Dan Rather put it, was "so nasty it would gag a buzzard." A glimpse emerged at the end of June, when four women whom EMILY's List was supporting for governor — Dianne Feinstein in California, Evelyn Murphy in Massachusetts, Barbara Roberts in Oregon, and, of course, Ann — attended a NOW convention in San Francisco and had breakfast together. One mentioned that she was the target of a whisper campaign regarding her sexuality. Suddenly it became clear: Dianne Feinstein was the target of a similar campaign. In Oregon, so was Barbara Roberts. The same for Evelyn Murphy in Massachusetts. Whether it was divorce or rumors of sleeping around or lesbianism, women candidates' private lives weren't private.

In fact, such attacks were par for the course when women ran for office. In addition to the full-bore Republican-campaign gay-baiting characterizations of Barbara Mikulski in 1986 as a "San Francisco–style liberal" who was "antimale," in 1988, Jolene Unsoeld had to fight off a brutal lesbian-baiting and Red-baiting campaign.

Ann had been targeted with whisper campaigns not just in this campaign but through much of her political career. When she had been county commissioner in Austin in the seventies, she would hear alternating rumors that she was a lesbian or a whore. Of course, with her wit and irony, Ann was terrific at deflecting such attacks. "I just want to know," she said, with perfect comedic timing, "if I'm a lesbian, why did I sleep with every man in Travis County?"

But the attacks continued. In the Democratic primary, they had come from Jim Mattox when he charged that Ann had "attempted to establish a purely female-type connection with these individuals, and I don't know whether that is normal or abnormal in the process of raising money." And now it was happening again in the general election. In July, two Christian activists proclaimed Richards "an honorary lesbian" for her opposition to state laws banning sodomy. According to the *Dallas Times-Herald,* both activists worked for organizations that were partially funded by her opponent Claytie Williams's campaign. A few weeks later, Republican radio ads in Houston began harping on Richards's support from gays and lesbians. And there was a widespread whisper campaign in churches and elsewhere suggesting that the breakup of Ann's marriage was tied to her support by the lesbian community. What's more, while her opponents tried to make Ann seem "other," they were painting a rosy picture of how Williams embodied the mythic past.

Theoretically, concerns about crime, drugs, education, and abortion were important to Texas voters—and, if that were actually the case, Ann should have done fine. But in this campaign, such issues became secondary to concerns over who best embodied Texas's mythic past. Having made his fortune in oil, cattle, and real estate in West Texas, Claytie Williams was a latter-day Jett Rink (the James Dean character in *Giant*), the multimillionaire West Texas wildcatter who owned no shoes (just cowboy boots), who had a swimming pool in the shape of an Aggie boot (the emblem of his alma mater, Texas A&M), and who cried real tears when he heard the Aggie War Hymn. Never having held an electoral office, Williams didn't have a record to run on, so he peppered the airwaves with TV spots that were powerfully evocative of the Old West. As he put it, "I am Bubba."

Being "Bubba" meant that Williams personified a rural Texas ideal found on the ranch, in the oil fields, and in the locker room, one that celebrates the virtues of toughness, self-reliance, and neighborliness, all clad in Marlboro Country cowboy imagery. Williams's message, then, was that he was just one of the guys — that there were no social barriers separating ranch hands from multimillionaires like him.

But Bubba had a dark side as well, and it was responsible for the crude level of political discourse of the campaign. Williams was quoted as saying he thought rape was just like the weather. "There's nothing you can do about it," he said, "so you might as well lie back and enjoy it." Allegations that he used to stage "honey hunts" — parties where hookers were hidden throughout his ranch — led to his disclosure that in his youth he had been "serviced," as he put it, by prostitutes. When Williams talked about Ann early in the campaign, he said he'd handle her just like he did his cattle: he would "head her, hoof her, and drag her through the dirt."

Needless to say, many voters, particularly women, were horrified by Williams, and Ann made plenty of hay about his "foot-in-mouth disease." "The only way Clayton Williams can win is if they keep him from being in the public eye," she told reporters, "if they can keep him from talking to the press, if they can take him only to places that are very staged, where he is not given access to the press."

But in July, Williams still led Ann by 8 points and was outspending her three to one. The Ann Richards who had wowed the nation at the 1988 Democratic Convention did not play nearly so well in Republican Texas, where the term *liberal* was political poison. As a result, she made a show of going dove hunting to prove how well she could handle a rifle, and, to the dismay of liberal supporters, took positions that were out of character with her past. While remaining pro-choice, she moved to the center, and even to the right, on gun control, the death penalty, and the flag-burning amendment, befuddling Austin liberals who had known her for years. "One doesn't know what her fundamental positions are," said *Texas Observer* publisher Ronnie Dugger, a longtime voice for progressives. "Ann needs to go back to her roots and come up with a program that has some originality."

Ann began running a series of attack ads portraying Williams as a shady wheeler-dealer whose companies had been sued 319 times — for sex discrimination, fraud, price-fixing, and misappropriation of natural gas, among other things. But nothing seemed to help. Williams's ad campaign, which linked Ann to Michael Dukakis, inmates on death row, the gay and lesbian caucus, and Jane Fonda, was so effective that an August poll showed Ann was trailing by 11 points. "You get the feeling that things are slipping away for Ann, that Claytie is keeping her off balance, that she's taking a lot of hits and not hitting back," said a Democratic county chairman.

Even the good old boys who had insisted that they be part of her campaign now kept their distance. "Richards is struggling with a double negative, so to speak," said George Christian, who had not long before been shaken by Ann's rapid rise. "She can't get out from under the voters' negative perceptions of her."

But loyalists insisted she was still in the running. "Ann's going to win," one of them said. "You just wait. Claytie's going to step on his dick real hard."

But was Claytie ready to oblige?

On October 11, at a joint appearance with Ann, Williams called her a liar and refused to shake hands with her, an act that was widely perceived as ungentlemanly. As the *New York Times* put it, the episode provided Ann's campaign an opening to exploit what was being called "the moron factor," suggesting that Williams was "a crude bumpkin unfit for high office."

Then, on October 29, a thirty-one-year-old Austin woman who had been raped testified that while the perpetrator was attacking her, he quoted Williams, saying that she should do as Claytie said and "relax and enjoy it." Outside the courtroom, the woman told reporters that rape is no laughing matter. "At no point, Mr. Williams, was I able to relax and enjoy it," she said. "Rape isn't a joke."

Now that the race was in the home stretch, Ann became more aggressive, but she shifted direction somewhat. "Ann has found her message," said one of her advisers. "Her environmental message is quite tough, and she has begun comparing Clayton Williams to Bill Clements [the unpopular Republican incumbent]. This is the politics of rich and poor."

Williams, after all, was presenting himself as Bubba — a man of the people — when in fact he was the owner of two private airplanes; a twin-engine helicopter; twelve ranches in Texas and Wyoming totaling 481,000 acres; a ten-thousand-square-foot mansion next to the Midland Polo Club with three carports, an indoor pool, guesthouse, and wine cellar; a bank; and an oil company. All told, Williams's wealth totaled $150 million.

Ann insisted that he release his taxes. Williams refused.

Ann's team repeated her demand. "It is an absolute necessity now that we know whether or not this man — who is one of the richest men in the United States — paid any income taxes like the rest of us," said Ann's press secretary, Bill Cryer. At one point, the campaign even sent a moving van to his office to pick up the tax returns. But Williams still said "no."

Meanwhile, at one campaign event after another, Ann vowed to take back the statehouse for the people of Texas. "It's really gonna be neat," she said. "And I want to invite you now . . . I want to make a date with you. We are going to meet on the Congress Avenue Bridge. And we are going to march up Congress Avenue and we are going to take back the capitol of Texas for the people of Texas." The crowd always went wild.

With less than two weeks before the election, a new Gallup poll showed that Williams's lead had shrunk to 5 points, putting Ann within striking distance. Even better, a Republican poll showed Ann and Williams in a dead heat, with 38 percent each. The tide was shifting. Williams had clearly lost all momentum. "It's coming down to who makes the last mistake," said Dennis Sheehan, chairman of the Tarrant County Democratic Committee in Fort Worth.

By the last weekend of the campaign, the tension was so great that I had to be part of it. So, on Friday, November 2, I flew out to Austin accompanied by Wendy Sherman, EMILY's List's new executive director, who had come on board after Rosa DeLauro had resigned to run for Congress herself. (Wendy had been Barbara Mikulski's campaign manager in 1986.)

After Wendy and I arrived in Austin, we met with Jane Hickie and got caught up on the events of the day. Earlier in the morning, speaking at the county courthouse in Longview, Texas, for the umpteenth time, Ann called on Williams yet again to release his tax returns. "You ought to be able to see whether or not someone who aspires to lead this state cares enough about government to pay his fair share in running the government," she said.

Then, early in the evening, Wendy and I piled into Jane Hickie's car to go out to dinner, when Jane got a phone call. We could hear only Jane's end, but it was apparent from her excited voice that something was happening. When she hung up, she shrieked with delight. This was the moment we had been waiting for. At the time, Williams was on a whistle-stop train trip from San Antonio to Houston, pledging, as usual, to improve the state's economy, veto any income tax, and fight a war on drugs. This time, though, Jane said, Williams had had an extra bourbon or two when a reporter asked him about his taxes. He parried the question as he had countless times before. "Yes, I've paid lots of income tax, lots, lots," he said.

But then, he slipped. "I'll tell you when I didn't pay any income tax was 1986, when our whole economy collapsed."

"What happened next was hysterical," said Jane Hickie. "The Williams campaign realized how bad this was. They put all the reporters on buses and trains and drove them to Mexico and tried to get everyone drunk."

But it was too late. Later that day, the Associated Press ran the story with the headline MILLIONAIRE CANDIDATE SAYS HE PAID NO INCOME TAX IN 1986.

And so it went, not just in Texas, but all across the country. OPEN MOUTH MAY CLOSE A DOOR IN TEXAS read the *Washington Post*. A NEW GAFFE BY CANDIDATE IN THE TEXAS GOVERNOR'S RACE headlined the *New York Times*.

Chuck McDonald, a spokesman for Ann, said that Texans never held Williams's great wealth against him, but the fact that someone that rich had paid no taxes was another matter entirely. "When the perception is that a man who lives far, far better than the average Texan didn't have to pay taxes one year, people do not see that as any way fair," McDonald said. McDonald described it as the final blow for Williams.

That last weekend before the election, President George H. W. Bush—he of the silver foot—ended up spending most of his time in Texas, making joint appearances with Williams, as if rushing to Claytie's rescue.

On Sunday, he and Williams sat down at a GOP phone bank when reporters persuaded Bush to set aside lists of registered Republican voters and instead pick a name at random from the phone book. "This is risky," Bush said. Then, he went ahead and dialed a number listed in the name of Malissa Johnson.

"Ms. Johnson. You won't believe this. This is the president of the

United States calling . . . I'm trying to get people out to vote . . .
Needless to say I hope you'll vote for our full ticket."

There was a pause while a woman responded.

"Will ya?" Bush told her. "I'm so pleased."

Sitting nearby, Williams broke into a grin and breathed a sigh of
relief.

But not much later, a reporter called the woman, who revealed
she thought Bush was an impostor. Moreover, she had told him only
that she would be voting — but not for whom.

In fact, she intended to vote for Ann Richards.

OUTSPENT AS IT WAS by Williams, the Richards team had unpaid
volunteers in almost all of Texas's 254 counties, and phone banks in
most of them. On Tuesday, November 6, that team put on a full-
court press to get out the vote. By noon, CBS exit polls showed Ann
ahead 52 to 48 percent. That was particularly good news given that
Ann's voters were said to be late voters.

According to the book *Storming the Statehouse,* by Celia Morris,
that afternoon Ann played bridge with her son Dan and some of his
friends, and then went to an AA meeting. At about 9:30 p.m., her
campaign received a call saying that Ann's supporters at the Hyatt
Regency Hotel in Austin were likely to riot unless she got to their
headquarters soon.

At first, Ann was somewhat reluctant to go. If she went, she
couldn't avoid being on national TV, possibly without knowing
what the outcome would be. And she refused to go under those
conditions.

"I'm not going until you tell me I'm going to win," Ann said. "Yes
or no? I need to know: am I going to win?"

Ann was addressing Nancy Clack, a political consultant on her
team who had been incessantly running projections on the returns.
She kept coming up with the same result: it was going to be close,
but Ann was ahead.

Now the pressure was on Nancy. Her projections gave Ann 50.03 percent of the vote. With the Libertarian candidate on course to take 3 percent, there appeared to be no way Williams could beat Ann. Nancy stuck her neck out.

"You're going to win," she said.

So, off they went to the Hyatt. Just as she was about to take the platform, Jane Hickie's cell phone rang. By this time, the results had become increasingly clear. Williams's campaign manager, Kenneth "Buddy" Barfield, was on the phone, saying that the candidate wanted to speak to Ann. That could mean only one thing: he was ready to concede.

By then, Ann was holding aloft a T-shirt that read A WOMAN'S PLACE IS IN THE DOME. The crowd was going crazy. She made her acceptance speech, then took the phone. Williams congratulated her and graciously made his concession.

At the time, Texas's macho culture notwithstanding, women had been elected mayor in five of the biggest cities in the state — Corpus Christie, Dallas, El Paso, Houston, and San Antonio. That said, for Ann to be elected governor was truly groundbreaking.

SWEET AS ANN'S TRIUMPH WAS, her election was just one of many victories that year. In the House, in September, Patsy Mink (D-HI) had won a special election. In November, our own Rosa DeLauro emerged victorious, as did Barbara-Rose Collins (D-MI), Joan Kelly Horn (D-MO), and Maxine Waters (D-CA). Along with Mink, that meant that five pro-choice Democratic women whom EMILY's List had endorsed had been elected. In addition, three vulnerable women in the House, Jill Long, Nita Lowey, and Jolene Unsoeld, all won reelection by handsome margins. Plus, there was Eleanor Holmes Norton (D-DC), a nonvoting delegate to the House from Washington, D.C. Four of the women who won — Mink, Collins, Waters, and Norton — were women of color. In the four short years since EMILY's List had started, the number of Democratic women in the House of Representatives had increased by two-thirds, to twenty women, the highest number in history.

Nor was Ann our only gubernatorial victory. Dianne Feinstein narrowly lost in California, but two other women were elected governor. One was Joan Finney in Kansas, who was anti-choice and so not endorsed by EMILY's List.

The other, Oregon's secretary of state Barbara Roberts, was the single parent of an autistic child and had not finished college. At the time, there was no requirement in Oregon that schools had to provide special education for autistic children, and Barbara thought that was wrong. So, Barbara set out to change the law. She told me how she thought lobbyists were supposed to buy legislators fancy meals to win over their votes — and she couldn't afford even a cup of coffee. But Barbara was an extremely persuasive woman and, thanks to her work, Oregon now offers education for children with special needs. Come November 1990, Oregon also ended up with a wonderfully compassionate governor.

All of which helped make 1990 a terrific year for EMILY's List. Together, as we often said, Ann Richards and Barbara Roberts would govern more people than all previous women governors elected in their own right combined.

And there were other gains. EMILY's List itself had grown to 3,527 members — up 74 percent in just two years. The total amount of money we raised had tripled the fund-raising of the previous cycle. It was now $2,782,000 — about half of which went to candidates and half to operations, including our growing political program.

A LITTLE MORE THAN two months after the election, on a windy but sunny Tuesday in January, Ann Richards made good on her pledge to take back the state capitol in Austin with a mass march up Congress Avenue to the seat of government itself. Because U.S. troops were massing to launch Operation Desert Storm and force Iraq out of Kuwait, the eyes of most of the world were focused on the Middle East, but this was a truly groundbreaking event I would not have missed for the world. Wendy Sherman, Judi Kanter, and I flew to Austin and joined up there with a dozen other EMILY's List friends from across the country.

When she was elected Texas treasurer, Ann had become the first woman elected statewide since the days of Gov. Miriam "Ma" Ferguson, who had been elected in 1925, following the impeachment and conviction of her husband, Gov. James "Pa" Ferguson. But Ma Ferguson, of course, was merely a proxy for her impeached husband. Ann had been elected in her own right. We had elected a prochoice Democratic woman to the highest seat of state government in Texas, and she was truly a woman of national stature.

Women all over the country felt that way. "I remember I was in the state senate," said Debbie Stabenow, host of an EMILY's List party at her home in Michigan, who was working her way up the political ladder. "And I said to a friend, 'Why don't we go to Ann's inaugural? This is really historic.' So, we paid some ungodly sum for a last-minute airline ticket, but it was worth it."

For me, the festivities began that morning with a reception for women leaders held in an office building on Congress Avenue, as about fifteen thousand people were massing to begin the march. What had started as a rhetorical device in Ann's campaign speeches had become a reality, and the inauguration was characterized by a spirit of inclusiveness. "This was a first ever in gubernatorial Inauguration Days," said San Antonio mayor Henry Cisneros, "to have the people, not units of command, but people gathered together in a parade led by the governor elect."

When the march finally crossed the Congress Avenue Bridge and reached our office building, it came to a temporary halt to allow us and the women leaders of Texas to take their place at the head of the line.

Leading the way was Ann, flanked by her daughter Cecile, her son-in-law Kirk Adams, and her son Dan Richards. Signs in the crowd read EAT YOUR HEART OUT, CLAYTIE and THE PEOPLE OF TEXAS ARE BACK. (The latter was a quote from Barbara Jordan in endorsing Ann.)

Ann's gestures of inclusion toward women were not merely symbolic. Indeed, even during the transition period before being sworn in, Ann appointed Lena Guerrero to be the first woman on the historically white, male Texas Railroad Commission, an extraordinarily powerful regulatory body that, its name notwithstanding, oversees the regulation of the oil and gas industry, gas utilities, and more. And the Guerrero appointment was just the first of many women and minorities to committees and boards and other bodies that had never heard a woman's voice before.

The evening of the inaugural, Judi Kanter and I were invited to a reception at the governor's mansion for big donors to the campaign. When we arrived, Judi and I joined the receiving line that wound its way through the living room of the mansion.

During the transition period between the election and the inauguration, Democratic funders and Texas special interests realized they had a big problem. They hadn't supported Ann in those final crucial months of the campaign. Suddenly, to get in her good graces, they started sending contributions to the new governor as fast as they could.

But Ann knew exactly what was going on. At the reception, Judi and I could see her standing by the fireplace, shaking hands and talking with her visitors. Then, Ann spied Judi and me across the room. "Well, Ellen Malcolm, how are you?" she called out loudly. The governor was sending a clear message to the special interests of Texas: she knew who her *real* friends were, and who just wanted to climb on the bandwagon.

WHEN WE RETURNED to Washington, we had an enormous amount of work to do preparing for that once-in-a-decade opportunity that redistricting offered us. After the census was completed, some experts projected that there would be more than one hundred competitive seats, roughly three times as many as usual. Fortunately, the additional funds we had raised allowed us both to target these extra seats and to expand our political operations.

We were also fortunate that new executive director Wendy Sherman had acquired an immense amount of experience in field and political operations, communications, and issue development in federal, state, and local campaigns, including as Barbara Mikulski's campaign manager in 1986. Wendy continued to be a chief informal adviser to Mikulski. As we worked throughout 1991, our meetings were frequently interrupted when the receptionist would come in and tell Wendy there was a call for her: "The senator is on the line." That meant Sen. Barbara Mikulski, of course. It *had* to be Mikulski, not just because Wendy had worked for her but because there was only one Democratic woman in the Senate.

And so it went. Every week or so, the receptionist would pop her head in a meeting and tell Wendy the senator was on the line.

Before long, I was fed up — not because Wendy kept getting calls from Barbara Mikulski but because EMILY's List had been around for three elections and there was still only one pro-choice Democratic woman in the United States Senate.

Finally, it happened yet again and this time I knew exactly what to say. "I want us to resolve here and now that after the next election, when someone comes in and says that the senator is on the line, we are going to be able to respond, 'Which one?'"

And that's exactly what we did.

Ten

NANTUCKET SLEIGH RIDE

IN EARLY 1991, I got a lunch invitation from Rep. Barbara Boxer (D-CA). At the time, California was in the unusual position of potentially having *both* senate seats up for grabs. Because Pete Wilson had become governor, a special election was scheduled for his seat, for which former San Francisco mayor Dianne Feinstein was a very strong candidate. In addition, campaign issues and health problems plagued Democratic senator Alan Cranston — and Barbara had her eye on his seat.

Over sandwiches in Barbara's congressional office, we talked about her senatorial prospects. I advised her that her highest priority was to travel the state and put together a solid fund-raising base.

After that, Barbara called me periodically, full of enthusiasm. "I spoke to the League of Women Voters in Orange County, and they were all excited," she told me. "I think they really liked me."

"Yeah," I quickly replied. "But did you ask them for any money?" We often laughed about my insistent emphasis on money — but I knew it was absolutely essential to finance the huge TV campaign that was necessary to win in California.

Meanwhile, the women EMILY's List had already elected were flexing their muscles in Congress, and we began to see the fruits of their labors. Women legislators were the driving force behind the Family and Medical Leave Act and health-related bills before

Congress such as Patsy Mink's Ovarian Cancer Research Act. Pat Schroeder, as chairwoman of a subcommittee on the Armed Services Committee, had been given credit for single-handedly making family issues—child care, education, equal pay—mainstream political issues.

Thanks to Barbara Mikulski's work in the Senate, the National Institutes of Health launched a major new women's health initiative. It was hard to believe, but until the mid-eighties, clinical studies for many critical diseases, such as heart disease and stroke, included only male subjects. Even though laws had been changed in the mid-eighties, no one paid attention until 1990, when Barbara, using her clout overseeing NIH appropriations, led an effort to create the NIH's Women's Health Research Initiative, which launched long-term studies on a variety of women's health issues, the largest effort of its kind ever conducted in the United States.

In addition, in the House, a formidable triumvirate sometimes known as DeLowsi—a reference to Rosa DeLauro (D-CT), Nita Lowey (D-NY), and Nancy Pelosi (D-CA)—had begun putting forth legislation to provide a safety net for women, families, and kids. Most notably, in 1991, they succeeded in getting funding for breast cancer research put into defense appropriations.

THEN, COMPLETELY OUT of the blue, something happened that had an enormous impact on EMILY's List. It reminded me of a term I learned during my summers on Nantucket referring to whalers who harpooned their gargantuan prey, which then took off at breakneck speed. When that happened, the sailors had no choice but to hold on for dear life on a "Nantucket sleigh ride" that was full of excitement and had the promise of a great reward. My sleigh ride was about to start.

On June 27, 1991, Supreme Court justice Thurgood Marshall, the first and only African American on the nation's highest court, announced his retirement. The strongest liberal voice on the bench, Marshall was not happy that it fell to President George H. W. Bush, a Republican, to name his replacement. But Marshall was eighty-two years old and in poor health. "The strenuous demands of Court work," he said, were "incompatible with my advancing age and medical condition."

The question of who would replace Marshall instantly became a heated Washington parlor game. Needless to say, there was enormous pressure on President Bush to appoint another black justice to the court. But Bush also now had an opportunity to put a conservative stamp on the court that could last for ages. *Roe v. Wade* was just one law of the land that could be threatened. All this was of great interest to me, but, initially at least, it wasn't clear that it would have much impact on EMILY's List.

Then, just four days after Marshall's announcement, on July 1, 1991, Bush named Clarence Thomas as his nominee. A graduate of Yale Law School, Thomas had been appointed in 1974 as assistant attorney general of Missouri under Missouri attorney general John Danforth. When Danforth became a United States senator, Thomas moved to Washington and became his legislative assistant. Then, Thomas served as assistant secretary of civil rights in the U.S. Department of Education and, in 1982, was appointed by Ronald Reagan to be chairman of the Equal Employment Opportunity Commission.

Over the next eight years, Thomas became a favorite of conservatives thanks to his vocal opposition to affirmative action. Instead of filing class-action suits against discrimination, the usual EEOC approach, Thomas aggressively promoted a doctrine of self-reliance, accompanied by heated rhetoric he used to eviscerate a number of black civil rights leaders.

Now, after announcing Thomas's nomination to the nation's highest court, President Bush said that "the fact that he is black and a minority had nothing to do with this." Of course, that was ludicrous. Even Republicans bemoaned Thomas's lack of qualifications. According to the book *Strange Justice*, by Jane Mayer and Jill Abramson, the most authoritative examination of Thomas's nomination, officials in Ed Meese's Justice Department during the Reagan administration had been highly critical of Thomas's legal scholarship. When Bush became president, White House counsel Boyden Gray drew up a list of potential Supreme Court nominees, and Thomas's name was not on it.

And with good reason. His experience as a jurist was minimal. In 1990, Bush had appointed Thomas to the U.S. Court of Appeals, but he had been on the bench for less than two years. He had never written a single constitutional opinion. He had never tried a single case before a jury. And just a year earlier, when Justice William Brennan resigned, President Bush's advisers warned him that the American Bar Association would almost certainly find Thomas unqualified for the Supreme Court.

But, thanks to the racial politics surrounding the nomination, all of that was of no consequence. As the descendant of American slaves, Thomas had ascended from rural poverty in a fabulous narrative that gave cover to Bush and the Republicans against charges that they were insensitive to minorities and the poor. Sen. Orrin Hatch (R-Utah) confirmed as much when he warned Democrats, "Anyone who takes him on on the subject of civil rights is taking on the grandson of a sharecropper."

None of that meant Thomas was immune from criticism, however. Knowing that the pastor of Thomas's church had compared abortion to the Holocaust, feminists protested that Thomas might vote against *Roe v. Wade*. But the Bush White House diffused such attacks by making certain that Thomas took no position whatsoever on abortion during the confirmation process. To do otherwise might jeopardize support from moderate Republican senators such as Arlen Specter (R-PA).

At the same time, the White House quietly began lobbying Democratic senators who might be swing votes for Thomas. Vice President Dan Quayle put in a call to his golfing buddy Sen. Alan Dixon (D-IL), who, it turned out, was up for reelection and likely to face a successful Republican businessman named Gary MacDougal. According to *Strange Justice,* Quayle told Dixon that a vote for Thomas would be valued highly by the White House. Later, when Dixon made his support for Thomas known, MacDougal found that much of the support he had from wealthy Republicans mysteriously began to vanish.

Meanwhile, Kenneth Duberstein, the former White House chief of staff under Reagan, was brought in to shepherd Thomas through the hearings before the Senate. Under Duberstein's aegis, Thomas underwent rigorous coaching sessions in the Old Executive Office Building, mock trials of a sort, to prepare him to be grilled by the Judiciary Committee. In addition, Thomas visited privately with more than sixty senators before the hearings got under way. Conventional wisdom had it that his confirmation would be a breeze.

What few people knew, however, was that in late July, a lawyer for the Federal Communications Commission in Washington named Gary Phillips got a call from an old friend of his from Yale Law School, Anita Hill, who had become a law professor in Oklahoma. When Phillips asked her what she thought about Clarence Thomas's nomination, he was stunned by what he heard. Hill had worked under Thomas at the EEOC and quit, she said, because he had sexually harassed her. According to *Strange Justice,* the last thing she wanted was "to get involved in a political three-ring circus," but Phillips thought she was struggling with whether or not to come forward. At this point, of course, almost no one was aware of her charges.

Instead, when Thomas's Senate hearings began, on September 10, the nation heard the moving story of his journey, which began in Pin Point, Georgia, where his mother made twenty dollars a week as a maid. It was, as Thomas put it, "a life far removed in space and time from this room, this day, this moment." Some of his testimony was the stuff of inspirational dreams, but for the most part Thomas came across as somewhat wooden and unresponsive.

MEANWHILE, AT EMILY'S LIST, Wendy Sherman had decided to join a media-consulting firm. Rather than replace her as executive director, I brought on Karin Johanson as political director, with an eye toward focusing on all the new open seats created by redistricting.

BOTH MEN AND WOMEN often ask me why it's important to have women in Congress. Why shouldn't they just vote for the best candidate? Ann Richards, with her record-setting appointment of women and minorities and her clear ability to lead Texas, provided one very clear reason. But an even more powerful and dramatic answer was about to materialize in the form of a fascinating national psychodrama played out on TV before tens of millions of mesmerized viewers.

It started drearily enough. For a full week, Clarence Thomas's testimony was a study in evasion and nonresponsiveness. The *New York Times* characterized it as "a blend of childhood reminiscences, quips, regurgitations of elementary constitutional theory, platitudes, and, in one instance, what Senator Biden called 'the most unartful dodge I ever heard.'" A federal appeals-court judge, John Minor Wisdom of New Orleans, called the hearings "the most unilluminating I've ever heard." In his late-night monologue, *Tonight Show* host Johnny Carson declared, "Last night's audience was influenced by Clarence Thomas. It's not that they didn't like the material. They just felt it was wrong to give an opinion on it."

Such criticism aside, Thomas's testimony was highly successful in one regard: it seemed likely he would be confirmed. THOMAS WINNING THROUGH EVASION; ELUSIVE NOMINEE SKIRTS OPPOSITION headlined the *Boston Globe*.

But then, on October 6, two days before the Senate was scheduled to vote on Thomas's confirmation, the floodgates burst open. National Public Radio's legal affairs correspondent, Nina Totenberg, reported that Anita Hill said Thomas had sexually harassed her when she was his special assistant at the EEOC. Hill, who had since become a professor at the University of Oklahoma College of Law, said that Thomas often talked to her about his sexual interests and sexual acts he had seen in pornographic films, that he asked her out socially many times, and that he graphically discussed his own sexual prowess and anatomy.

The full Senate was eager to proceed with the vote on Thomas's confirmation, on Tuesday, October 8. Barbara Mikulski urged a delay in the vote, but she was the only Democratic woman in the Senate.

So, on the House side, on the very day the vote for confirmation was scheduled to take place, seven Democratic women — Reps. Barbara Boxer, Pat Schroeder, Louise Slaughter, Patsy Mink, Nita Lowey, and Jolene Unsoeld, and delegate Eleanor Holmes Norton — marched over to the Senate side of the Capitol to tell the senators that they needed to take Hill's allegations seriously and reopen the hearings. Because Senate rules forbade outsiders from attending party caucuses, they were barred from meeting with the senators. Senate majority leader George Mitchell met separately with the women and told them to call Democratic senators who were supporting the Thomas nomination.

The Judiciary Committee, under the chairmanship of Democratic senator Joe Biden, was doing everything it could to minimize the issue. But, thanks to Totenberg's story, it was impossible to put the toothpaste back in the tube. The entire nation — or at least the female half — was up in arms. So, on that day, the Senate Judiciary Committee bowed to the political pressure and reopened Thomas's confirmation hearings.

Prior to this moment, the phrase "sexual harassment" had been an obscure term that was rarely uttered in public. As such, it served as the ultimate gender-biased Rorschach test. Sexual harassment? The vast majority of men didn't have a clue. Remember, to a large extent we were still living in a *Mad Men* world in which men were the bosses and women were valiantly struggling to build their careers. Men thought nothing of patting a woman on her backside, commenting in great detail about her appearance, or making an off-color joke at her expense. So, as Anita Hill's story unfolded, millions of men wore puzzled looks or dismissed the charges as frat-boy high jinks. Most men had little or no understanding of the sexual politics behind male–female employer–employee relationships. They didn't want to. It wasn't in their interest.

But to tens of millions of American working women, this was an earth-shattering moment. At one time or another, they had witnessed or experienced sexual harassment and had suffered in silence to protect their jobs and reputations. Women who had never told anyone they were victims of sexual harassment suddenly came out of the proverbial closet. In elevators, at watercoolers, in carpools, and over the phone they tiptoed into the great unknown, speaking the truth about inappropriate behavior foisted upon them in their workplace.

Working women discussed the many tactics they would use to avoid further harassment and/or retaliation by their bosses. They spoke of trying to diffuse the situation with humor. They told of attempts to keep from being alone with the harasser. They talked about trying to get transferred within their company or looking for a new job. They talked about losing promotions or being fired because they wouldn't "put out."

Anita Hill's courage in publicly making her charges unleashed years of pent-up rage in many working women. They believed Clarence Thomas's behavior was horrible and that he should not be on the United States Supreme Court. Quite simply, they believed Anita Hill, and they wanted to know where the women senators were who could defend her. "People kept asking me, where were all the women in the Senate?" Mikulski said. "Why aren't there women on the Judiciary Committee? And when I explained, there are only two of us in the entire Senate, they said, 'Wow!' "

The next day, October 9, the EMILY's List phone started ringing off the hook with queries, mostly from Los Angeles, asking for membership information. Soon enough, Judy Krantz called from L.A. and explained precisely why. When she got the *Los Angeles Times* that morning, the first thing she saw was the front page of the style section, featuring a big picture of Gov. Ann Richards smiling and saying, "I would not be the governor of Texas today if it were not for Ellen Malcolm and EMILY's List."

The timing couldn't possibly have been better. Just when women all over the country were rising up in anger, Ann Richards's ringing endorsement boosted our national credibility. If there were more women in office, perpetrators of sexual harassment would be punished, not elevated to the United States Supreme Court.

Meanwhile, the hearings reconvened. On Thursday, October 11, Anita Hill herself finally testified, stating under oath that Clarence Thomas had sexually harassed her, first while he employed her at the Department of Education, and later at the EEOC. She said that Thomas asked her out on dates many times while he employed her, and when she demurred he began introducing graphic sexual subjects at work, such as women having sex with animals, and discussed films depicting rape or group sex. Hill testified that Thomas repeatedly discussed his sexual expertise and his own anatomy. In what would become perhaps the most famous detail of her testimony, according to Hill, Thomas once pointed to a can of Coca-Cola on his desk and asked, "Who put a pubic hair on my Coke?" A narrative about racial politics had been transformed into one about sexual politics. The entire country was riveted.

Hour after hour, senators grilled Hill, questioned her veracity, and made her out to be a spurned woman seeking revenge. Here you had a woman who was a graduate of Yale Law School and had gone on to become a soft-spoken, understated law professor, and the Republicans were characterizing her as promiscuous and delusional. "A little bit nutty and a little bit slutty," became the Republican mantra.*

This was not exactly the kind of colloquy one expected from

---

* The phrase was coined by David Brock, a writer who had been designated by the right to destroy Hill's reputation and burnish Clarence Thomas's. In a book called *Blinded by the Right,* Brock later recanted and described how he had assembled "virtually every derogatory and often contradictory allegation" about Hill to help Thomas's nomination.

what was supposedly the nation's most dignified deliberative body. But it was precisely the sort of discourse that kept tens of millions of Americans glued to their TV sets hour after hour, day after day, watching a national spectacle unfold the likes of which had not been played out since Watergate, seventeen years earlier.

Yet, as time went on, one thing became increasingly clear: explosive as Anita Hill's charges against Clarence Thomas were, the way her allegations were handled — or, rather, mishandled — was equally astonishing. Her Republican critics asserted that Hill had come forth at the eleventh hour to smear Clarence Thomas. But that wasn't true. In fact, a report of Hill's allegations had been made known to the Senate Judiciary staff three months earlier, in July.

Hill was not asked to come forward to the Senate Judiciary Committee until September 12. She gave the FBI sworn affidavits and cooperated with its investigation, which ended on September 25. Yet, somehow or other, even then the Judiciary Committee, chaired by Senator Biden, did not see fit to question Thomas about the allegations. It was not Biden's finest hour.

In other words, here, in the midst of an intensely bitter, partisan political battle, were serious allegations that would provide the Democrats with powerful ammunition against Thomas's confirmation, yet the Democratic Party had been giving Anita Hill the runaround for weeks.

And why was that?

The first answer was quite simple: they were men. Out of the fourteen members of the Senate Judiciary Committee, there was not one single woman. And the men in the United States Senate, Democrats as well Republicans, were as clueless about sexual harassment as other men throughout the United States. Gender trumped partisanship.

As reported in *Strange Justice,* by Jane Mayer and Jill Abramson, Sen. Howard Metzenbaum, when first apprised of the allegations, unwittingly explained the situation. "If that's sexual harassment," he said, "half the senators on Capitol Hill could be accused." In other words, many of them had harassed women, and, to the extent that they had, this was the last subject they wanted to discuss.

There were other factors as well that kept the Democrats from acting. Four years earlier, they had waged a bloody, unprecedented battle that had killed President Reagan's nomination of Robert Bork to the Supreme Court, and many simply didn't have the stomach to go through that again.

But the paralysis in the Senate was a pathetic response to the tens of millions of women who were in an uproar, made worse by the fact that the victim of the harassment was being publicly vilified for coming forward. The gender divide had never been more explicit. It was there for all to see. On October 15, 1991, Thomas was confirmed in the Senate, 52–48, the narrowest margin of approval in more than one hundred years. When the roll was called in the Senate, you heard the word *Mister* fifty-two times before the first woman was called. One can only imagine the result if women had made up more than 2 percent of the United States Senate. As if to make the point abundantly clear, the National Women's Political Caucus took out an ad in the *New York Times* featuring a drawing of fourteen women senators facing one lone man: Clarence Thomas. The tagline was "What if . . ."

EVEN THOUGH THE NOMINATION was over and Clarence Thomas was headed for the high court, the fallout from the hearings had just begun. "The Anita Hill experience was such a defining, historic moment," said Hillary Clinton, whose husband was just starting his 1992 presidential campaign at the time. "I had gone to law school with Clarence Thomas, and, though I didn't know him well, he seemed surly to me even then. And here was this incredibly put-together woman, and at great cost to herself she comes forward and ends up being treated like some kind of heretic. 'Burn her at the stake!' I was appalled. I was so proud when Barbara Boxer and the whole crew marched over."

Suddenly, women all over the country saw the need to elect

women. EMILY's List's phones rang off the hook. A woman from Chicago called to tell me I absolutely had to come there to tell her and friends — women professionals at large Chicago businesses — how to help elect women senators. I was on the phone so much that I lost my voice. We quickly ran out of EMILY's List brochures. One reporter after another called me, too. Among them was a woman journalist who, by the end of the conversation, had dropped all pretense of being a disinterested, neutral observer. "Oh, my God! I was so furious!" she shouted. "I couldn't believe how they treated Anita. Where were the women?"

Clearly, something extraordinary was happening. Judi Kanter, Betsy Crone, and I agreed we needed to regroup to maximize our marketing efforts — immediately. Betsy moved our direct-mail recruitment schedule into high gear. She and our writer, Paul Bennett, met with me at the office to discuss a new solicitation package keyed to the Thomas-Hill hearings. The whirlwind continued. On my way to the West Coast, I had a brunch in Chicago with about twenty women professionals who held significant positions in accounting, law, and other businesses. "I was a child of the sixties," one of them said. "I protested the war in Vietnam and went to all the marches. But when I got married and my career took off, I put aside politics. Now, I have to get back involved."

This was our market — women who had marched against the war in Vietnam, who had lived through and appreciated the political passions of the tumultuous sixties. These women, now in their forties and fifties, had significant financial wherewithal and untapped political power. They had careers and appreciated how difficult it was to balance work and family. Now, the disgraceful way Anita Hill was treated and the disregard for women in the workplace had reignited their political passions. EMILY's List was exactly what they had been looking for: a way to elect women like themselves to office. EMILY's List was a way for them to make their own decisions and to know they were really making a difference.

Judi Kanter was organizing events fast and furiously. I flew to Houston for an event with Sherry Merfish, who had hosted our party three years earlier with Ann Richards and Mary Landrieu. This time, things were different. When Sherry picked me up at the airport, she was agog at what had been going on. There was so much demand for the event that she had run out of invitations. More were being printed. The phone wouldn't stop ringing.

Sherry feared the hosts would be upset. "We're going to have about three times as many people as I told them," she told me. Fortunately, the weather was good, so we used the hosts' backyard.

Later that afternoon, when I arrived at the event, I was flabbergasted to see the registration line all the way down the sidewalk. Even more astonishing, when Sherry escorted me into the backyard, I was met by three TV camera crews. On occasion, print reporters had shown up at our events in the past. But TV? Never.

When Sherry had hosted the same event three years earlier, about 40 people attended, roughly 25 of whom actually wrote checks. We had considered that a real success. But this time, 450 people paid $100 at the door to join EMILY's List and left eager to tell their friends all about us. Moreover, TV news crews spread the word far and wide.

The reverberations from Hill's testimony were still being felt all over the country. In November, just after the Thomas hearings, I joined nine hundred women — four hundred of them state legislators from all over the country — at a dinner at the Hotel Del Coronado in San Diego, at which Anita Hill was the guest of honor. When she arrived, she was greeted like a rock star, with one woman after another, Republican and Democrat alike, climbing atop her chair, waving a pink napkin and chanting, "An-i-i-i-ta, An-i-i-i-ta."

By now we knew that a seismic change was under way that would be felt for years to come. We quickly moved to harness the energy from all over the country and transform it into political power. It was clear we would raise record amounts of money. Sexual harassment wasn't the only issue. Polling data showed that women were angry that Congress didn't have a clue about what was going on in the lives of working women and their families. The women polled thought elected women would understand the economic pressures of daily life and would be tough fighters on those issues. EMILY's List was flooded with phone calls from celebrities and working women alike, all wanting to elect more women.

Women candidates became the symbol of the change voters were yearning for, the end of "politics as usual." In Washington State, Patty Murray, a citizen lobbyist for environmental and educational issues with little political experience, ran for the United States Senate. In California, Dianne Feinstein declared, "Two percent is fine for low-fat milk, but not for women in the United States Senate." Barbara Boxer ran for the second Senate seat in California. In Pennsylvania, Lynn Yeakel took on Republican senator Arlen Specter, who, having accused Anita Hill of "flat-out perjury," had emerged as one of the archvillains of the Thomas-Hill hearings. In the House of Representatives, a banking scandal had created a "throw the bums out" ethos. And, of course, virtually all the "bums" were men.

Some of the women candidates didn't appear to stand a chance. While I was traveling that fall, back in Washington, EMILY's List political director Karin Johanson had a visit at the office from an impressive forty-four-year-old African American woman who had been a federal prosecutor and state representative in Illinois. As the Cook County recorder of deeds in Chicago, she had been enraged by Sen. Alan Dixon's vote for Clarence Thomas and had decided to challenge the incumbent Democrat. She was completely focused on how badly the process had failed. "She was unbelievably charming," said Karin. "She was really smart. Terrifically impressive. She had a great smile. She had gone to Yale Law School. She was having a fine career there in Cook County. All I could think was, 'My God, I really want to support her.'"

There was just one problem. When it came to the question of viability, this impressive newcomer, named Carol Moseley Braun, by every measurement we knew, wasn't even in the ballpark. Challenging an incumbent in one's own party was problematic enough, but Sen. Alan Dixon had won twenty-nine consecutive elections. And even if you argued that Dixon's vote for Clarence Thomas made him vulnerable to the rising anti-incumbent tide, he already had a strong challenger for the upcoming Democratic primary: a wealthy liberal Chicago lawyer named Al Hofeld, who boasted a hotshot campaign adviser in David Axelrod and was starting off with $5 million in his war chest.

By contrast, Moseley Braun had no money to speak of—just commitments of roughly $80,000, and those were commitments, not real money. The primary was scheduled for March 17, 1992, just four months away, and Moseley Braun had started her campaign, if you could call it that, way, way too late. Even if, by some miracle, she got traction, she and Hofeld would split the anti-incumbent vote, thereby ensuring Dixon's reelection. This one was too much of a long shot for us. Karin gave Moseley Braun what encouragement she could, but we couldn't put our resources behind her.

MEANWHILE, I WAS OFF to New York for another fund-raising trip, and on November 19, I got a message from a producer at CBS's *60 Minutes* named Patti Hassler. *60 Minutes,* of course, was and remains the most successful newsmagazine in television history. I called back immediately.

At the time, one of the stars of the program was Mike Wallace, the hard-boiled investigative reporter known for his devastatingly tough interviews of the famous and infamous. Was that what I was in for? At first, I was somewhat guarded. But soon enough, I figured out that Patti was sincerely interested in how the Thomas-Hill hearings would impact women in politics. Fascinated by every detail, she had an especially curious mind. She probed me about EMILY's List until she knew as much about it as I did. Soon, I decided the more the *60 Minutes* folks knew about us, the better the piece would be. When I told her how excited women were at our events in Chi-

cago, Houston, and elsewhere, Patti insisted on seeing the phenomenon firsthand. We had four big events coming up in February in
Florida: in Miami, West Palm Beach, Tampa, and Tallahassee. CBS
would be with us at all of them. I quickly called Judi Kanter to get
her to orchestrate things, to make sure everything ran smoothly.

EXHILARATING AS THESE DAYS were, we had a few rough moments as well. One of our most difficult decisions had to do with the
upcoming Senate battle in New York, where two strong pro-choice
women, Geraldine Ferraro and Elizabeth Holtzman, were vying for
the Democratic nomination to run against the Republican incumbent, Al D'Amato.

We had known something like this would happen sooner or
later. Earlier that year, in fact, I raised precisely this issue before the
EMILY's List steering committee, saying that I expected we would
soon have races with more than one pro-choice woman candidate.
But now it was not a mere hypothetical. "Here we had two well-
respected, iconic feminist figures," said Karin Johanson. "How were
we going to figure this out?"

Ferraro, of course, had been the first woman vice presidential
nominee in either party, and had a sterling record in Congress.
Holtzman had been elected to Congress when she was just thirty-
one, and had served as district attorney in Brooklyn. In 1980, she
had lost the New York senatorial race to D'Amato by just 1 percent,
and, with the Clarence Thomas backlash, this time she might well
be propelled over the top.

To make things more complicated, a number of people on the
steering committee were close to Gerry Ferraro. My Common
Cause "good government" training taught me to make sure all decisions we made — especially ones that might be controversial — were
done in a fair and proper way. So, I quickly put together a small
group of independent advisers who had no relationship whatsoever with either Ferraro or Holtzman. Their mission was to find
out who would be the strongest candidate to win the Senate race.

That meant deciding who would be the strongest candidate to win the general election, not just the primary. During their evaluation process, they would help us define the decision-making process for future candidates. To that end, they met with the two candidates and with their campaign managers. They examined the campaign strategies and funding. And they commissioned a statewide poll.

AS 1992 CONTINUED, I realized that had the Clarence Thomas hearings happened five years earlier, there was no way we would have been prepared as an organization to take advantage of the fall-out. But we had by now resolved critical questions about how EM-ILY's List would operate and had built an infrastructure to help us recruit members and raise money. Even though we had a staff of only seven, we were positioned to take advantage of the unusual political opportunities the 1992 election offered because of redis-tricting. For me personally, I had learned a great deal about how to lead a small but significant political organization. I was no lon-ger as terrified of public speaking and had acquired some skills in fund-raising and media interviews. Best of all, I was as excited as any member to turn the disaster of the Thomas-Hill hearings into a political victory for women. I was fired up, and there would be no stopping me. By January, our membership was skyrocketing, and the primary season was just getting under way.

So, when we arrived in Florida in early February, I was opti-mistic about the *60 Minutes* piece. We had been traveling with the show's crew to the Tallahassee home of Florida education commis-sioner Betty Castor, with four women running for Congress. Scores of women were there. I had given my presentation time and time again, but on this occasion, Patti was there with the *60 Minutes* camera crew and I was wearing a wireless mike.

Feeling impassioned, I went into my talk about how the Clarence Thomas hearings were fueling our movement. "I can tell you the anger and the fury and the energy that came out of those hearings is going to turn into support for women candidates," I said, "and we're going to have a record number of women elected in 1992 because of what happened in there. Yes!"

As I spoke, my eyes panned the room, much as the CBS camera was doing. One woman after another was cheering or giving me a thumbs-up sign. There was an incredible new kind of political energy. The women all had that look of infectious excitement in their eyes. Finally, standing in the corner, was Patti Hassler. When I saw she had the same look of exhilaration as the other women, I thought, "Whew, this story is going to be okay."

Two days later, I was flying across the state with Patti, and I tried to get her to tell me what would happen when the *60 Minutes* piece on us aired. But Patti was the consummate professional. She wouldn't bite.

I tried another tack. "Tell me about other stories you've done," I said.

"Well," she said, "if you're the watercooler story on a Monday morning, you will get a big response."

How big? I asked. Ten phone calls or ten thousand?

"Very big," she said. "The show airs Sunday night, and if everyone is talking about it Monday morning, you can have a pretty phenomenal response."

Now, that gave me something to think about. As soon as we landed, I called the can-do Betsy Crone and told her we needed to make special arrangements for all the incoming calls we would get.

"I'll see if I can get another phone line," Betsy said.

"No!" I said. "You don't understand. We may get thousands of calls. Tens of thousands."

Betsy went off to see what we could do.

ON FEBRUARY 14, we finally decided that Gerry Ferraro was the strongest candidate in the New York senate race. In large part, that was because Ferraro had a bigger following upstate than Holtzman, whose support did not extend much beyond New York City. In addition, thanks to her big national following, Ferraro was in a much stronger position to raise funds in the rest of the country.

I believe we came to our decision in a fair and rational manner, but that didn't make Elizabeth Holtzman happy. "What bothers me most about your rush to judgment is, it harkens back to the same machine politics all of us have fought long and hard to defeat," she said in a letter to me.

We knew we would get grief for it — whichever way we decided — and, of course, we were right. Holtzman's campaign manager, Ed O'Malley, compared EMILY's List to the Soviet Politburo. According to him, I just gave marching orders and everyone did as I said. "You might as well put a cigar in Ellen Malcolm's mouth," he said.

Well, that's politics. Our mission was to add as many pro-choice Democratic women to office as possible. For us *not* to have made a decision would have been fainthearted. And that meant that at various times, we at EMILY's List were hard-nosed pols. You have to be to win.

So, I took it all with a grain of salt and sent O'Malley a box of White Owl cigars. I made sure that a short note accompanied the cigars. "Have one on me!" it said.

MEANWHILE, THE PRESIDENTIAL PRIMARIES were under way. President Bush had emerged from the Gulf War with an astoundingly high approval rating of 89 percent. But since then, the country had plunged into a recession so deep that he had been in freefall in the polls. As a result, 1992 became one of those rare election cycles in which even an incumbent president was vulnerable. A number of high-profile Democrats, including New York governor Mario Cuomo and Sen. Al Gore, had declined to seek the nomination, however, meaning that the field was dominated by former Massachusetts senator Paul Tsongas, California governor Jerry Brown, Sen. Bob Kerrey of Nebraska, and a relatively unknown governor of Arkansas named Bill Clinton.

I was particularly interested in Clinton, because I knew Hillary was a key figure in the Children's Defense Fund, the American Bar Association, and the Children's Health Insurance Project — in short, a woman who was a leader in her own right. If her husband won, she would likely be a very different kind of First Lady, indeed, the first woman professional to live in the White House. In fact, I'd heard so many incredibly great things about Hillary that that alone made me curious about the Clinton candidacy.

On Sunday, January 26, 1992, just before the Iowa caucuses, Bill and Hillary went on *60 Minutes* to rebut allegations by a woman named Gennifer Flowers, an Arkansas state employee, TV reporter, and part-time cabaret singer, that she had had an extramarital affair with Bill. It turned out to be a memorable and highly successful damage-control operation. This was essentially the first time Hillary had been introduced to America. I think she succeeded in confronting an embarrassing and awkward situation while retaining great personal strength and dignity. "There isn't a person watching this who would feel comfortable sitting on this couch detailing everything that went on in their life or their marriage," she said. "And I think it's real dangerous if we can't have some zone of privacy." Then came the kicker. "You know, I'm not sitting here — some little woman standing by my man like Tammy Wynette. I'm sitting here because I love him, and I respect him, and I honor what he's been through and what we've been through together. And you know, if that's not enough for people, then heck, don't vote for him."

Hillary ended up taking flak from fans of Tammy Wynette, the country singer. But I felt she held her ground with dignity.

That year, no fewer than thirteen women were running for the Senate in hopes of keeping Barbara Mikulski company. The reason was simple. "That all-white, male Judiciary Committee was for so many women a dramatic visual image of a reality most women have been aware of for years," said Ruth Mandel, a professor at Rutgers University and director of the Eagleton Center for Women in Politics. "Women all over have known what it is to confront all-male boards of governors, all-male banking commissions, all-male agencies, all male boards, usually behind closed doors, but this time it was a face-to-face dramatic confrontation, and televised."

Some of those women senatorial candidates were real long shots — none more so than Carol Moseley Braun in Illinois, whose candidacy was explicitly tied to the Clarence Thomas–Anita Hill episode. "If the Senate had done its job right from the start, we all would have been spared the mess," Carol said. "And who were these guys anyway? Where were the women, the minorities and the regular working people?"

Moseley Braun had gone ahead with her poorly financed non-campaign and had not made much headway. "Talk about your underdogs," said Tony Podesta, a college friend of Moseley Braun's who became her political consultant. "I couldn't even find a professional fund-raiser who she could pay to work for her." As a result, donors who intended to support her sometimes were not even called. Moseley Braun did not have enough money to travel throughout the state. For the most part, she was an afterthought — if that. Chicago's Rotary Club, for example, invited both the incumbent Democratic senator, Alan Dixon, and his challenger Al Hofeld to speak before their luncheons — but not Moseley Braun.

As the campaign unfolded on the Chicago airwaves, Hofeld went after Dixon hammer and tongs, while Dixon returned the favor with TV ads lambasting Hofeld as a rich trial lawyer. The result was that Carol emerged unscathed, but still no one knew who she was. She had virtually no media campaign.

When Moseley Braun made public appearances, she was able to generate enormous enthusiasm. But she had little else going for her. And to the extent she got free media in the local papers, it wasn't all favorable. Her staff was rife with internal conflicts. Campaign manager Kgosie Matthews had never managed a campaign before, and, according to Alton Miller, Moseley Braun's press secretary, the effort on Moseley Braun's behalf was "so screwed up that I couldn't be a part of it anymore." Miller and two other top campaign aides quit.

Initially, the polls showed Dixon with nearly a two-to-one lead over Carol. At first, Hofeld trailed badly. But then, as his media blitz unfolded, in February, he gained ground. First, he was neck and neck with Carol, and finally, in early March, he edged into second place. One poll showed Carol a full 30 points behind Dixon with just two weeks left in the campaign. It would take a miracle for her to win.

Then, on Sunday, March 8, just nine days before the election, something began to happen. For one thing, a new poll showed that Dixon had slipped to 38 points and Moseley Braun had moved to 28, just 10 points behind him. If one believed the wildly fluctuating polls, she had somehow chopped 20 points off a 30-point gap she had had just a few days earlier.

Until this point, voters knew very little about Moseley Braun except that, as the *Chicago Tribune* put it, "She's not Hofeld or Dixon, and apparently that's enough to make many of them want to vote for her." Outspent more than twenty to one by Hofeld, even at this late date Carol had not run one single television ad — unheard of for a serious senatorial candidate.

Then, later that same Sunday, at a televised debate between the three Democratic contenders, Carol came off as the strongest, in part because Dixon and Hofeld were beating up on each other. Thousands of people disliked those nasty men who couldn't seem to be nice, and Carol was the epitome of likability.

The next day, Monday, March 9, with just eight days left before the primary, thanks to Carol's fine debate performance, enough last-minute funds poured into her campaign that she could finally begin running TV ads. "People do like to be with somebody who's perceived as having a chance," said Sue Purrington, the Chicago director of the National Organization for Women. "I think the poll and the debate have fired up her campaign. We've already gotten a lot more calls here about 'How can I help? I saw her last night.'"

Civil rights leader Jesse Jackson, a one-time presidential candidate and a powerful voice in the black community, jumped in himself with last-minute radio ads for Moseley Braun, a move that was certain to shore up the black vote. But there were just eight days left until the election.

Then, on Wednesday the 11th, Gloria Steinem, the cofounder of *Ms.* magazine and, as the *Chicago Sun-Times* put it, "a figure of mythic proportions in the women's movement," put together a breakfast in Chicago that poured $25,000 into Carol's campaign. "An investment in social justice is the best investment you can make," she said. Then, Steinem took out her checkbook and wrote a personal check of $1,000 to Moseley Braun's campaign.

There were now six days left to Election Day.

Then, Gloria called me. Something was happening in Chicago, she said. EMILY's List should support Carol. You shouldn't write her off, she told me.

Frankly, I was astonished. Gloria knew we were serious about our commitment to getting as much bang for our buck for EMILY's List members. That meant not taking fliers on lost causes. Much as we liked Carol, she had never seemed viable. She had recently been down by as many as 30 points. She had raised only roughly $150,000 at this point, whereas Hofeld had already spent more than $3 million of his own money. And now, with just five days left, Hofeld was digging into his bankroll for at least another million.

"But Gloria, she doesn't even have a campaign," I said. "She's not even on television, so I doubt anyone knows she's running."

But Gloria pushed back. "I'm telling you something is really happening," she said. "There is an incredible amount of energy out there. Women are still furious about the Thomas-Hill hearings, and they want Carol to win."

Normally, EMILY's List either recommended a candidate or didn't; it was all or nothing. The only possible case we could make for Carol's success was that, with Hofeld and her in a seesaw battle for the anti-incumbent vote, all that anger and energy might tip in her favor. Hofeld's massive ad campaign had inflicted enormous damage on Dixon, but it didn't make voters like Hofeld. What if

Carol was the real beneficiary? Jesse Jackson was now a vocal supporter of Carol, so the black vote might turn out in force for her. And, of course, as I had been telling everyone, this was the year of the woman. With Gloria Steinem, the National Organization for Women, and, of course, us on the case, perhaps we could work a miracle.

And so, at the last minute, with no time left for a fund-raising mailing, we endorsed Carol and gave her the maximum $5,000 direct contribution.

ON THE EVENING OF MARCH 17, as the returns rolled in, it became clear that the quid pro quo that Senator Dixon had made with Dan Quayle had exploded in his face. Assuming that his vote for Clarence Thomas would ensure smooth sailing in November, Dixon hadn't given a moment's thought to the political power of women. Now he was paying for it. At 11:30 p.m., Bob Schieffer of CBS News reported, "Senator Alan Dixon is in the fight of his political life with a Cook County courthouse official named Carol Braun who did not even run a television ad until last week."

In the end, primary night in Illinois hit the political world like an earthquake. The impossible had happened. Carol Moseley Braun, a black woman with little statewide name recognition, a minuscule ad budget, and next to no television advertising, had somehow conquered a previously undefeated longtime incumbent senator in the Democratic primary.

Dixon's vote for Clarence Thomas had inadvertently lit the fuse to forge a powerful new coalition between urban black voters and suburban white women angered by Dixon's betrayal. According to the exit polls, Moseley Braun got 82 percent of the black vote, and, among suburban white women who opposed Thomas's confirmation, she outpolled Dixon by a factor of more than 6 to 1.

Moreover, this phenomenon was not limited to Carol. Rather, she had become, as the *Chicago Tribune* put it, "head of a de facto feminine slate" that swept into office in the state. Something dramatic had happened. "The whole day was a women's day," said Chicago alderman Richard Mell. "It was something that just started growing, was sort of a word-of-mouth thing. It wasn't anything formal."

Many people simply voted for every woman they could find on the ticket: women local officials, women legislative candidates, even women judges. According to Judy Erwin, who won the Democratic nomination for Illinois state senator from Lincoln Park, "When there was an opening and not an established, well-liked incumbent, the advantage seemed to go to women."

Three Democratic women sought nominations to the Metropolitan Sanitary District Board. All of them won. In the races for Cook County judgeships, forty women sought nominations, and three-quarters of them won.

It was only March, and Illinois was just the first state to hold federal primaries. But if Carol Moseley Braun could defeat an incumbent senator in the primary, the sky was the limit. The Year of the Woman had officially begun.

Eleven

THE YEAR OF THE WOMAN

TICK, TICK, TICK, TICK.

In the world of television news, no iconography is more instantly recognizable than the inexorable sweeping second hand accompanied by the sound of the ticking clock that opens CBS's *60 Minutes*.

Tick, tick, tick, tick.

In the five months since I first met *60 Minutes* producer Patti Hassler, I had been through the full gamut of emotions. Initially, wary of the program's reputation for hard-hitting interviews, I had been cautious.

Tick, tick, tick, tick.

Then, once Patti won me over, I was desperately eager to make sure she had all the relevant materials necessary to understand what made EMILY's List work. For months, I did everything possible to figure out how to shape the story, how to get her to understand all the pieces of what we do. I flooded Patti with information. I showed her how we reinvented fund-raising. I explained the $100 checks as a marketing initiative. I invited her to our events. I made key members of our staff available.

Tick, tick, tick, tick.

We had done all we could to prepare for the show. As usual, Betsy came through. She found a company that would split our calls into sixteen different lines with a recorded message asking people to leave their names and addresses. We also prepared a massive direct-mail recruitment campaign — eight hundred thousand pieces or so — which would be ready to go the day after the CBS piece was broadcast. We even shipped boxes of the letters with West Coast addresses to Judi Kanter in San Francisco; she could take them to the post office there so they would be delivered faster.

Tick, tick, tick, tick.

At 7:00 p.m. on Sunday, March 22, 1992, I was at my home in northwest Washington with Karin Johanson, waiting for *60 Minutes* to begin. My mind was in overdrive. What if they changed their minds and did one of those Mike Wallace–style hit jobs? What if women had lost interest in the Thomas-Hill hearings? What if nobody cared? I knew my fears were baseless . . . but still.

Either way, this time the ticktock was for real. It ticked for EMILY's List and for me.

Before I knew it, CBS's Morley Safer was on the air, saying how outraged hundreds of thousands of women were about the Clarence Thomas hearings. Then, the camera cut to me at one of our Florida events. "I can tell you that the anger and the energy and the fury that came out of those hearings is going to turn into support for women candidates, and we're going to have a record number of women elected in 1992 because of what happened there," I said. The camera then panned across the room to show scores of attentive and excited women.

Then, back to Morley Safer. "The first thing you should know about EMILY's List is that there is no Emily," he said. "EMILY is one of those overly cute acronyms standing for 'Early Money Is Like Yeast . . . It makes the dough rise.'"

I winced. *Oh, no,* I thought. Safer was more an avuncular type than an attack dog, but maybe he was coming after us just the same.

Then, he paused briefly, and continued. "But there the cuteness ends." At that moment, I breathed a sigh of relief. After that one light dig, I knew we would be okay.

And, in fact, what happened next on *60 Minutes* was beyond our wildest dreams. I had told Patti how we had used the $100 figure as a marketing tool to get women to write bigger checks, and there it was, being explained in the voiceover as, on film, one woman after another wrote checks. CBS even squeezed in a last-minute bit about our endorsement of Carol Moseley Braun.

And the next day, we were the watercooler story all over America!

We received sixteen thousand phone calls over the next week from people wanting membership information. Returns started pouring in from our massive direct-mail campaign. Our computer system needed to be upgraded and our check-processing system overhauled to manage the incredible growth.

It was exhilarating and exhausting. Finally, I pulled our small staff together to review how we were doing. "Look," I said. "I don't think it will ever be like this again in my lifetime. The members are signing up, and the political opportunities are mushrooming. So, we have to do all we can now to take advantage of this moment. Let's work together to do the best we can. And we'll clean up the rest after the election!"

Over the next year, my life was transformed. This was my fifteen minutes of fame. When my planes landed, TV crews greeted me at the airports as if I were a celebrity. In Houston, I was introduced as "the woman who changed the face of American politics." I had never paid much attention to fashion, but suddenly, what I wore was important. I had to learn how to be in the media spotlight.

Most importantly, all of this was being converted into political action. EMILY's List had closed the 1990 elections with more than three thousand members, but, in the immediate aftermath of the *60 Minutes* segment, we shot up to eleven thousand members, with no end in sight. Women who had never participated in politics before were signing up. Women who once contributed to the Democratic Party were writing checks to us instead—because their own party had ignored them.

Now, with the rising tide of anti-incumbent, pro-woman senti-

ment, women were real players. "For the first time in my political
life, and that's twenty-five years, I feel that we are empowered," said
Hedy Ratner, an EMILY's List fund-raiser.

Just a few months earlier, Carol Moseley Braun had been ignored
by the Democratic establishment. But now that she had beaten Senator Dixon in the primary race, when she went to Washington for
money and backing, she was hailed as a conquering heroine. Democratic power brokers including Senate majority leader George J.
Mitchell of Maine and Massachusetts senator Edward M. Kennedy
were lining up to meet with her.

By this time, all six of the women we endorsed for the 1992 Senate races were considered to have a serious shot — quite an improve-
ment from two years earlier, when the number was zero. Barbara
Mikulski was thrilled with what was happening. "This is truly history in the making," she said at one fund-raising event. "When you
look at the women standing here on this platform, you know that
the New World Order is here!"

As for the House of Representatives, the election was insane.
Because of redistricting, there would probably be one hundred
new members of Congress — roughly triple the number in an average election. At EMILY's List, women would just call us up out of
the blue, saying, "I want to run for Congress." Even then, I knew I
would look back on this year and tell young friends a generation
later, "Well, it all started in 1992."

One of the most interesting races was for a Pennsylvania Senate seat. Lynn Yeakel, a fund-raiser for Women's Way, had been inspired to run for the U.S. Senate after watching Arlen Specter, her
state's Republican senator, browbeat Anita Hill. Yeakel was not by
any means a one-issue candidate, but she repeatedly made the Clarence Thomas–Anita Hill testimony a focus of her campaign. "To
have ninety-eight men and two women in the Senate in 1992 is just
simply not right," she told CNN. "This is a year when women are
going to step forward — or have stepped forward — and the voters
are going to see us as agents of change, as real hope for fixing a

system of government that is not working well, and that's why this will be a year when record numbers of women will be elected to Congress."

Another virtual unknown, Lynn had no electoral experience whatsoever and faced a strong opponent in the Democratic primary in Lt. Gov. Mark Singel. In March, about a month before the primary, polls showed that only 1 percent of the Democrats would vote for her. In any another year, EMILY's List would have decided that Lynn did not pass the viability test.

But in the wake of Carol's victory in Illinois, it was clear that these were unusual times. The dynamics had changed. We could help shape events — not merely respond to them. So, we endorsed Lynn, as did a host of women's organizations. The money started flowing into her campaign, and on April 21, Lynn hit a grand slam, winning the endorsement of Pennsylvania's four largest newspapers.

That same day, a poll showed Lynn closing in on Singel, running just 5 points behind him in a three-way race. Then, just three days later, she took the lead, 32 percent to 29 percent, and she never looked back. When the returns rolled in on April 28, once again it was as if the Year of the Woman had been officially sanctified by the electorate. Lynn Yeakel had come from nowhere to win the Democratic nomination for the Senate in Pennsylvania.

Similarly, in Washington State, Patty Murray, who had become a state senator, was running for the U.S. Senate seat being vacated by Brock Adams, a Democrat who dropped his reelection campaign after eight women came forward accusing him of various acts of sexual misconduct. In a normal year, Patty would not have passed our viability test either.

It would be an understatement to say that Patty was not wired into the political establishment. Her father, who had once been the manager of a five-and-dime store, had been forced to apply for welfare when he was stricken with multiple sclerosis. Patty herself had been a preschool teacher and later taught a parenting class at a community college. Then, in the eighties, when Patty was an activist for environmental and educational issues, an unaccommodating state senator inadvertently provided her with the words that would launch a political career. "You're just a mom in tennis shoes," he said. "Go home. You can't make a difference."

To Patty, those were fighting words, so she transformed that patronizing put-down into her rallying cry. First, she was elected to the school board; then, the state senate. Now, after less than four years as a state senator, she was moving into the big leagues. "Look out world!" she told a cheering crowd. "Us moms in tennis shoes are going to take over!"

Patty's campaign took up residence in a building built by Lou Graham, a flamboyant nineteenth-century madam who ran what was said to be "the best little whorehouse in Seattle," and Patty proudly acknowledged its origins. "It's a hundred years later," she told a reporter, "and women still are having to sell themselves."

Of course, it's a cliché now for candidates to run for office as Washington outsiders. But Patty *really* meant it. In fact, it was such a fundamental part of her identity that, initially at least, Patty was wary of *us* at EMILY's List as Washington insiders.

That was fine with us. But, in view of Moseley Braun's and Yeakel's primary victories, we still had to determine whether the "mom in tennis shoes" was on a path to victory. So, we did our own independent poll in Washington State, and we realized people were sick to death of politics as usual. When Patty talked about helping families, voters said, "That's what I want. She gets what my life is like. I'm voting for her."

In late April, we took the unusual step of endorsing Patty at a time when Democratic governor Booth Gardner and other Democrats were still considering running. "This is an incredible boost," Patty told a reporter. "It sends a message to everyone else that this is a serious campaign. They make it clear they don't endorse long shots."

The very next day, whether it was because of our endorsement or other factors, Governor Gardner made his decision not to enter the race against Patty. I've always known that, in some measure, politics is about perception. We had said Patty wasn't a long shot. Now, it was so.

MEANWHILE, IN CALIFORNIA, there was the possibility that for the first time in history, one state might be represented by two women senators. Former San Francisco mayor Dianne Feinstein had been so impressive in her governor's race two years before that she was already considered the front-runner to win the special election to fill the two remaining years of the seat that then-governor Pete Wilson had vacated after winning the California gubernatorial election.

But Barbara Boxer had a tougher election in front of her. For the Democratic nomination alone, she faced two strong candidates who had powerful advantages over her. In Lt. Gov. Leo McCarthy, Barbara had a foe who was far better known statewide, a huge plus in a state as big as California with its exorbitantly expensive media markets. And in Rep. Mel Levine, she faced a five-term congressman from Los Angeles who was part of the so-called Waxman-Berman Machine, named for two liberal Democrats from Los Angeles who controlled the most potent political operation in Southern California. This was the liberal California version of the old boys' network, and it meant Levine would have no problems whatsoever when it came to money. In fact, he had assembled roughly $4 million for the primary, which would enable him to spend more than $500,000 each week in television ads. That was a number Barbara could not compete with.

California being California, a state with massive political clout, this primary was seen as a real test of the Year of the Woman thesis. As the *New York Times* put it, the outcome would "help answer the question of whether the Clarence Thomas confirmation hearings, and the anger that many women expressed at the all-male Senate Judiciary Committee, have made women significantly stronger as candidates."

Barbara was hurt by the fact that, as a member of Congress, she was tied to a banking scandal in which it was revealed that the House of Representatives allowed members, including Boxer, to overdraw their House checking account without the risk of being penalized. But she compensated for that by telling the effective story that she was campaigning in the same outfit she wore when she and six other women members of Congress tried to enter the Senate to ask the all-male Judiciary Committee to hear out Anita Hill.

"Do you know what they told us?" she asked her audience. "Do you know why they wouldn't open their doors to us? They said they don't let strangers in." Then, she paused for effect.

"And if they consider members of Congress to be strangers, what do you suppose they think about the rest of you?"

This strategy was effective, and, even after massive advertising by her opponents, by April 30 she had taken the lead by 26 percent to 24 for McCarthy and 21 for Levine. A key to her success, especially given Levine's initial lead in fund-raising, was that she was able to collect $4 million in donations from fifty-two thousand donors, more than two-thirds of whom were women.

On July 2, Boxer cruised to a surprisingly easy triumph. Dianne Feinstein achieved victory on the same ballot, and the two women Democratic senatorial nominees immediately went on an extraordinary statewide victory tour to kick off their campaigns for the general election. At the first stop on their tour, at the Burbank airport, they basked in the glory for a few moments and each gave a speech. "It is a little awesome," said Dianne. "I looked over at Barbara, and I had just a feeling in the pit of my stomach. There is something more than just politics happening. There really is. It is a kind of phenomenon."

The Boxer-Feinstein duo was such an amazing spectacle that I was asked about it repeatedly by reporters, and it prompted my favorite speech line ever. "If one more reporter asks me if California is ready to elect two Bay Area, Jewish, liberal, right-handed, dark-haired, lipstick-wearing women," I said, "the answer is, 'HELL YES!'" In the five months since the *60 Minutes* piece, there were hundreds of articles decreeing 1992 as the Year of the Woman — not to mention hundreds of articles mentioning EMILY's List, more than in our entire history to that point combined. Our membership had already grown from some three thousand before the Clarence Thomas hearings to more than fifteen thousand. We were on the road to raising $5 million for candidates, which would make us the single biggest fund-raiser for candidates in the United States. And the election was still four months away.

The 1992 Democratic National Convention opened on July 13 in Madison Square Garden to nominate Bill Clinton and Al Gore as its presidential and vice presidential candidates. A few months earlier, I had met a wonderful woman in New York named Judy Loeb Goldfein, who was eager to work with us. So, I asked Judi Kanter to team up with her on an event designed to highlight the seven Democratic women running for the Senate. We called the event Faces of Change United States Senate (FOCUS), and we booked the biggest ballroom at the New York Hilton in Midtown Manhattan.

When the DNC convention opened, women played an unprecedented role. On the first day of the convention, July 13, *People* magazine threw a luncheon honoring me, attended by opera singer Beverly Sills, TV star Marlo Thomas, ABC's Barbara Walters, and many more celebrities. Hillary Clinton joined Tipper Gore and our candidates for the Senate as we all celebrated an extraordinary year of triumph for women.

The next day, Tuesday, July 14, was time for FOCUS, where 450 people attended the $1,000-per-person VIP reception we hosted in honor of our seven senatorial candidates. Governors Ann Richards and Barbara Roberts spoke, as did Barbara Mikulski. "For six years I've been the only Democratic woman in the U.S. Senate," Barbara said. "But thank God that is over, because help is on the way."

Next, we all moved into the main ballroom, where more than three thousand men and women were standing. After being introduced by Glenn Close, I conjured up, with the help of a twenty-foot movie screen, a melodrama of sorts, introducing our heroine, Anita Hill; our villain, Clarence Thomas; one powerful image that "burned itself onto our political psyche: the picture of the all-male Senate Judiciary Committee"; and finally, with the soundtrack from *The Magnificent Seven* playing, our "Magnificent Seven" senatorial candidates—Barbara Boxer and Dianne Feinstein of California, Carol Moseley Braun of Illinois, Josie Heath of Colorado, Geraldine Ferraro of New York, Patty Murray of Washington State, and Lynn Yeakel of Pennsylvania. "In November," I said, "when the final speech is given, and the final call is made, and the final ad is run, and the final vote is counted, I think the world will be different. For you and I have done something extraordinary tonight. You and I have made the world a different place."

Everyone was having a fabulous time, cheering, crying, and laughing, and it was stunning to hear people come over and ask, "Can I give you another contribution?" In all, we raised more than $750,000 that night, making it by far the most successful fundraiser for women in history.

TWO DAYS LATER, on July 16, 1992, Bill Clinton formally won the Democratic nomination at a convention that was a real breath of fresh air. This time around, it was well-planned, relatively gaffe-free, and united—especially for a party that had a history of being divided. There was an air of optimism that was exemplified by the campaign theme song, Fleetwood Mac's "Don't Stop (Thinking About Tomorrow)." With Bill and Hillary Clinton and Al and Tipper Gore, the Democrats had at the top of the ticket two couples from the sixties generation in which the women were accomplished professionals in their own right.

Anita Hill's spirit pervaded the convention. Along with EMILY's List, the National Women's Political Caucus and the Women's Campaign Fund played an unprecedented role, in large measure because there were so many women candidates — strong women, real leaders, who belonged on the national stage. On the convention floor, Patty Murray's team handed out red PATTY MURRAY FOR SENATE shoelaces, a creation from Randy Murray, Patty's fifteen-year-old son, that worked perfectly with her slogan about being "just a mom in tennis shoes."

As my fifteen minutes of fame ticked on, Jimmy Carter invited me to build a house with him for Habitat for Humanity. Then, three weeks later, as the Clinton-Gore campaign was getting under way, I was invited to join Hillary at the American Bar Association convention in San Francisco, where Anita Hill was scheduled to receive an award.

Even though she was consumed by her husband's presidential campaign, Hillary still followed the Anita Hill saga closely. We all stayed at the same hotel. First, I met privately with Anita and was able to thank her for all she had done to help EMILY's List. Then, finally, I met Hillary, who asked me about all of the Senate campaigns, not as a wife and prospective First Lady but as a savvy political operative inquiring in great detail about what was happening in all of our campaigns. I was impressed. Hillary was a real trailblazing activist for women in law and was going to be no ordinary First Lady. I'd heard so many great things about her. I was beginning to understand why.

ON SEPTEMBER 15, the primaries yielded spectacular results for us, with only one major disappointment — the Democratic primary battle in New York. Geraldine Ferraro, whom we had endorsed, had been leading by 14 points just three weeks before the primary. But

Elizabeth Holtzman went after Gerry with no holds barred, focusing on the Ferraros' tax problems and legal problems, even suggesting that Gerry's husband had ties to the Mafia, and the result was a brawl between two women that ended up helping the leading male candidate, in this case New York State attorney general Robert Abrams. When the votes were counted, it was clear that the attacks certainly didn't help Holtzman, who finished last with just 13 percent. But they also wounded Gerry so badly that Abrams squeaked by her and won by 1 percent.

In Washington, Patty Murray won her primary but now had a tough battle against her Republican opponent, Rep. Rod Chandler, a former TV anchorman and five-term congressman who mocked Patty by carrying around tennis shoes. "He was so convinced this ditzy woman in tennis shoes was not going to beat him," said Karin Johanson. "He was completely disdainful of her."

Patty's great strength was her truly disarming ability to address basic human needs in a straightforward manner that contrasted sharply with Chandler's conventional Washingtonspeak. "She'd say, 'I know that when we discuss health care, there's not going to be anyone at the table for Washington's families. There will be somebody for the doctors and somebody for the insurance companies, but if you want someone for Washington's families, that's me.' And she did it in a terrifically articulate way," said Karin.

Chandler relentlessly attacked Patty as a risky, inexperienced candidate. By mid-October, he had pulled into a dead heat with Patty and was outspending her by more than two to one. But finally his condescension backfired.

On October 14, as the second televised debate between the two was drawing to an end, Patty criticized Chandler for ignoring the concerns of families and instead voting for a congressional pay raise. Rather than deliver a conventional closing statement, Chandler responded to her critique by singing the refrain from the Roger Miller hit "Dang Me," a country ditty about an unrepentant philanderer who leaves his wife and child: "Dang me, dang me / They oughta take a rope and hang me."

When he finished, a hush fell over the audience, which just sat there in horrified, stunned silence. Then, Patty gave her terse rebuttal. "That's just the kind of attitude that got me into this race, Rod," she said.

There was nothing more to say. Nor was there any need.

ON ELECTION DAY, Tuesday, November 3, I stayed after work at our offices with the staff, Judy Lichtman, Betsy Crone, Joanne Howes, Marie Bass, Karin Johanson, and several others as the election returns came in. There was still considerable apprehension, but, whatever the outcome, it looked like there could not have been a more celebratory moment for the Founding Mothers. A year earlier, before the Thomas-Hill hearings, we had three thousand members. Since then, our membership had exploded to twenty-four thousand. And we had no fewer than twenty-nine newcomers running in the House and six in the Senate on election night. EMILY's List members had contributed more than $6 million to candidates and more than $4 million to help us run the organization. We had become the nation's biggest funder of federal campaigns. All of that would have been absolutely unimaginable a year earlier.

If you're a whaler, I suppose the worst part of a wild, wild Nantucket sleigh ride must be the moment when your quarry gets away. Now, as the returns came in, we would find out whether or not we got our whale.

Because EMILY's List was now a huge part of the story, reporters kept calling me throughout the evening, and Deborah Davis Hicks, our press secretary, sent me one phone interview after another as I watched the returns from a fourteen-inch portable TV across from my desk. But because California and Washington State were three time zones behind ours, results would not come in until the polls closed there at 8:00 p.m. Pacific time — 11:00 p.m. Eastern time.

In regard to the presidential race, there was every reason to be upbeat. The Clinton-Gore mantra, "It's the economy, stupid," had played very well indeed, and when the polls closed, at 8:00 p.m. in most states, electoral votes rolled in for Clinton across the northern part of the Eastern Seaboard — New York, New Jersey, Connecticut, Massachusetts, Maine, and more. At 9:00, Dan Rather announced that Clinton already had secured 238 electoral votes, with just 32 more needed to win the White House. California alone would put him over the top, and Clinton was heavily favored there.

Then, in Illinois, with the first senate race to be called in the entire country — a spectacular victory, with Carol Moseley Braun winning by 10 points and becoming the first African American woman elected to the United States Senate — we were already making history.

With so many races, there were bound to be some disappointments. In Pennsylvania, Arlen Specter had bombarded the airwaves with ads portraying Lynn Yeakel as an elitist blue blood whose father had voted against the Civil Rights Act of 1964 while in Congress, and played up her minister's criticism of the Israeli government so as to turn her into an anti-Semite. Never having campaigned before, Lynn didn't quite know how to handle the constant barrage of attacks, and Specter won by just over 2 points.

Then, at 11:00 p.m., I was in my office ending a phone interview with a reporter when I looked up at Dan Rather on the TV. "The polls have closed on the West Coast," he said. "CBS projects that Patty Murray will be the new senator from Washington State. CBS projects that Dianne Feinstein has won her Senate race in California. CBS projects that Barbara Boxer will be the new senator from California."

Down the hall, I heard my friends and colleagues in the reception area cheering and the sound of champagne corks popping. I ran out and hugged them. "Wow! Did you ever think? Hooray for EMILY's List!" It was one of the best moments of my life. We had done it!

Counting Carol Moseley Braun's victory, we had added four women to the United States Senate. And in the House, no fewer than twenty-one new Democratic women supported by EMILY's List were elected. That number was particularly remarkable when you remember that there were only twelve Democratic women in the entire House just three elections earlier. That meant there would be thirty-five Democratic women serving in the new Congress — almost triple the number in 1986. American politics would never be the same.

EMILY's List had grown from a tiny, marginal group of women to a major political fund-raising operation. And, of course, we now had Hillary Rodham Clinton in the White House, a First Lady who we believed would be a new national leader for issues important to women and families.

Twelve

REVERSAL OF FORTUNE

IF THERE WAS EVER a time to celebrate, this was it.

And so we did, again and again. One of the first occasions took place on November 10, when Dianne Feinstein, having won a special election, was sworn in so she could take office right away. I watched from the Senate balcony as she took the oath of office. The audience cheered, and the new senator turned and hugged three of her sisters in the Senate: Sen. Barbara Mikulski and senators-elect Barbara Boxer and Carol Moseley Braun. Later that afternoon, Dianne invited about seventy-five Californians and national women leaders to celebrate with her at a Capitol Hill restaurant, La Colline.

I walked into the restaurant with my special guest, Anita Hill, just as Barbara Mikulski was speaking. I'd seen Barbara be pithy, very funny, and knowledgeable. But this time, I saw a very different woman—choked up, with tears in her eyes. In that moment, I was overwhelmed with the enormity of how challenging, difficult, and lonely it must have been for her, the only Democratic woman senator in the ultimate old boys' club, the United States Senate, for six years. Most of us were celebrating the fact that we now had more women senators to truly represent us. But Barbara got to celebrate having a new kind of Senate with new colleagues. Perhaps best of all, they were "girlfriends," as Barbara put it.

When Barbara finished speaking, I walked with Anita Hill to

the front of the room. As soon as the crowd spied her, everyone broke into cheers. I was able to introduce her to the women she had helped elect to the Senate. Anita and Carol Moseley Braun were clearly excited to meet one another, and they shyly exchanged telephone numbers. It was extraordinary for me to see these two powerful African American women hug and thank each other, each knowing they had made history in 1992.

A few weeks later, in January, EMILY's List kicked off the Clinton inaugural with an astounding party for more than forty-two hundred people at the Washington Hilton, filling the hotel's two gigantic ballrooms and two additional rooms to capacity. Barbara took the podium to explain exactly how elated she was that she was no longer the only Democratic woman in the Senate. "Some women spend their whole lives waiting for Prince Charming to come," she said. "I've spent my life waiting for Dianne and Patty and Barbara and Carol."

After Barbara finished, the other new senators spoke — Barbara Boxer, Dianne Feinstein, Carol Moseley Braun, and Patty Murray — and finally I took the podium, with the idea of painting a picture of what we had accomplished. I called all the congresswomen by name to the stage. I had practiced my speech over and over, but I had not bothered to actually read the names. As I read them off, one by one, even I became overwhelmed by how many there were. There had been a fundamental shift in women's political power. It was awe-inspiring. Every man and woman in that room felt this incredibly fierce pride and awe. Women had finally taken that gigantic step forward.

And it really was gigantic. Prior to the 1992 elections, Patty Murray's preferred mode of transport had involved tennis shoes. Now she traveled on Air Force One with Bill Clinton. Not long before, Carol Moseley Braun had been a recorder of deeds. Now, both she and Dianne Feinstein sat on the Senate Judiciary Committee — the very same heretofore all-male committee of Clarence Thomas infamy. The powerful House Appropriations Committee now had *five*

Democratic women, including Rosa DeLauro of Connecticut, Nita Lowey of New York, and Nancy Pelosi of California — who stalked the halls of Congress striking the fear into the hearts of the political old guard that at any moment they might put forth legislation to provide a safety net for women, families, and kids.

More to the point, because we had a Democrat in the White House and the Democrats controlled both houses of Congress, the women could actually get things done — like legislation.

That became abundantly clear on Friday, February 5, 1993, just two weeks into the Clinton administration, when Bill Clinton delivered on one of his campaign promises and signed into law his very first piece of legislation as president. It was such a warm winter day — fifty-seven degrees! — that I got to sit outdoors in the Rose Garden at the White House with Judy Lichtman, one of our Founding Mothers and head of the Women's Legal Defense Fund; Vice President Al Gore, Sen. Edward Kennedy, and Senate majority leader George Mitchell, among others. We watched as President Clinton signed the Family and Medical Leave Act, the landmark labor bill giving workers unpaid time off to care for newborns or ailing relatives.

Working with Rep. Patricia Schroeder (D-CO), Judy had been battling to make the bill law for a full six years. It had actually passed both houses of Congress — only to be vetoed twice by President George H. W. Bush. Here you had the Republican Party — the party of "family values" — killing a bill that would have protected workers who had to stay home and take care of their families.

The bill had become an issue in the presidential campaign in September 1992, when Bush vetoed it for a second time. Bill Clinton said that if he were president, he would have signed it. Now, he was doing precisely that.

After Clinton's remarks in the Rose Garden, Judy spoke. "This is a very poignant moment, and my heart is very full," she said. "There is no better example of the kind of change the American people sought when they voted in November than the swift and historic signing of this legislation. The first bill signed into law by President Clinton opens a new era, a time when we get beyond the rhetoric and provide real support and economic security for American families who so urgently need it."

I beamed from my seat as my old friend stood with the president, receiving the victory and attention she deserved.

IN OTHER WAYS, HOWEVER, the Clinton administration was off to a dreadful start. Having made a campaign promise to allow openly gay men and women to serve in the military, the president encountered so much opposition to this issue upon taking office that he was forced to implement a policy known as "Don't Ask, Don't Tell," a compromise that left everyone unhappy. When it was revealed that his first two appointments for attorney general, Zoe Baird and Kimba Wood, had hired illegal immigrants as household help, both nominations were withdrawn. And, in their wake, more than one thousand other presidential appointments became the subject of intensified scrutiny, thereby creating logjams in the chaotic new administration.

In addition, of course, there was Hillary. As First Lady, she had taken on what was perceived to be a ceremonial role representing a womanly American ideal. But, by playing a substantive policy role in her husband's administration instead, she was now challenging America's assumptions about gender. At times, Hillary was even referred to as copresident, and, as a result, she became a primary target.

In fact, a number of First Ladies, from Abigail Adams to Eleanor Roosevelt, had played vital roles in their husbands' administrations. But that had been less true of recent First Ladies Barbara Bush and Nancy Reagan. So, from day one in the White House, the right demonized Hillary as a horrific Lady Macbeth–like figure on every topic imaginable — from the personal to the political, from the most trivial cosmetic faux pas to allegations about sexual infidelity and murder. "She was terrifying," said right-wing literary agent Lucianne Goldberg. "She was pushy, she was humorless. She couldn't get her hair figured out. There were so just many things about Hillary we didn't like."

As a result, the haters began searching high and low for any hint of impropriety that could be inflated into a massive scandal. One such episode came to the fore on May 19, 1993, when the Clinton administration fired seven employees in the White House Travel Office for various financial improprieties. Critics charged that this was part of a cunning scheme concocted by Hillary so she could shift the White House travel business to friends, including a third cousin of Bill Clinton's.

Neither the murky financial machinations in the Travel Office nor the firing of its staffers was the stuff of international import, and both matters were clouded by ambiguities and contradictory accounts. Nevertheless, before you knew it, no fewer than six major investigations — by the FBI, the Department of Justice, the White House, the General Accounting Office, the House Government Reform and Oversight Committee, and the Whitewater Independent Counsel — had been launched, and they would continue for more than six years. A molehill of a story had been transformed into an Everest-size mountain.

Moreover, these investigations weren't really about ethics or justice or truth. In the end, both Hillary and Bill were exonerated, but that didn't matter. What was really going on was that the right had learned how to seize control of the investigative process and the media and how to use them as political weapons.

Among its first targets was Vince Foster, a personal friend of the Clintons and colleague of Hillary's at the Rose Law Firm in Arkansas, who had become White House counsel and who sometimes discussed the travel-office affair with Hillary. On June 17, 1993, an editorial in the *Wall Street Journal* titled "Who Is Vince Foster?" proclaimed that the most "disturbing" thing about the administration was "its carelessness about following the law." Over the next month, the *Journal* painted the Clinton White House, Hillary, and her colleagues from the Rose Law Firm as some sort of corrupt cabal, with Foster being one of the ringleaders. The attacks were unrelenting, and Foster, a Little Rock lawyer who was deeply concerned about his reputation, couldn't take it. On July 20, he was found dead in Fort Marcy Park in Virginia. He had committed suicide.

After he died, a note was found in Foster's briefcase that may have shed some light on his state of mind. In part, it read as follows:

> I made mistakes from ignorance, inexperience and overwork.
> I did not knowingly violate any law or standard of conduct . . .
> The public will never believe the innocence of the Clintons and their loyal staff.
> The WSJ editors lie without consequence.
> I was not meant for the job or the spotlight of public life in Washington. Here ruining people is sport.

Far from ending the episode, however, Foster's tragic death only whetted the media's appetite for scandal. Right-wing conspiracy theories were rampant. On his Christian Broadcasting Network show, Rev. Pat Robertson suggested that Foster had been murdered and the Clinton administration had covered it up. Similarly, Rush Limbaugh, the right-wing talk-radio host whose show reached many millions each week, said that Foster may have died in a secret Clinton safe house and the corpse placed in Fort Marcy Park to make it look like a suicide. If the Clintons had murdered Foster, what were they trying to cover up? Had Foster been having an affair with Hillary? Had he killed himself because she ended it?

The Clintons had not been in the White House for even six months, but the war against them was being waged on many fronts.

FOR EMILY'S LIST, THIS was the beginning of a new stage in our development. Having just come off our first magical post-census "two" year, we confronted the fact that that was an event that wouldn't happen again for another ten years. Faced with a paucity of opportunities — maybe thirty or forty open congressional seats per cycle — we had to make the most of every single opportunity.

It was not going to be easy. For the first time in our history, the Democratic Party was in the White House, but midterm elections historically meant a reversal of fortune for the party in power. In addition, a new kind of vitriolic partisanship had begun to poison Washington. We had a new First Lady in the White House who supported EMILY's List and had become a primary target for the right. Plus, in 1992, women had won a disproportionately high number of close races, and it followed that these same women would be potentially vulnerable as they came up for reelection.

To counter these problems, we came up with a two-pronged strategy that signified an evolution for EMILY's List. First, we had to build from scratch a full-service political operation — something the Democratic Party should have been doing all along — that involved recruiting a new generation of women candidates, training them and their staffs, creating job banks, and providing seasoned professionals to advise them every step of the way so that they knew how to raise money, how to deal with the media, how to withstand political attacks, and more.

Second, given that we now were doing more to help women's campaigns, we needed to raise more money. That meant raising our fund-raising efforts to a whole new level, and going after the big donors as we never had before. I had wanted to undertake a political program like this for several years, and I had hired Rosa DeLauro, who had since departed for Congress, as executive director precisely for that reason. But it was not until the windfall of 1992 that we had the resources to really implement such an initiative.

This new direction was not exempt from criticism. Betsy Crone was concerned that fund-raising for EMILY's List might have an impact on the money we could raise for individual candidates — that we had been so successful in large part because members knew their money was going directly to candidates they had chosen. But I felt we could show members that helping candidates build stronger campaigns would then allow the candidate contributions to have more of an impact. It was essential to have a more hands-on approach to these races if we were to maximize our chances in each and every race. In virtually every election cycle, some of our candidates were political neophytes who had to put together a very complicated, sophisticated campaign in the glare of the media spotlight, and I knew we could be much more effective if they were better trained. This was especially true of House races, which were the entry-level campaigns for federal races. I often likened House campaigns to high school football. The team knew the rules and the gist of the game, but both staff and candidate were often in it for the first time. Occasionally, they'd be doing a great job only to make a colossal mistake, like the high school sophomore who grabs the football and starts running in the wrong direction!

The good news was that EMILY's List finally had enough resources to launch such a program. In our early years, our goal had been just to get people to come to our events and start writing $100 checks. But we now had twenty-four thousand men and women who had already written checks and had experienced great victories with us. Many of these were wealthy women whose resources had never been tapped.

Judi Kanter had ably demonstrated that taking on staff who reflected the life experiences of our members was very effective. Her "peer-to-peer" fund-raising reassured potential members, and she became a wonderful sounding board for me in understanding how to reach new women and how to excite them about giving more. So, we decided to expand the staff to build a major-donor program.

One person who would be terrific at raising more money from these new donors, I felt, was Judy Loeb Goldfein, who had been indispensable in pulling together our FOCUS event at the New York Hilton the previous year. So, Judi Kanter and I asked her and Sherry Merfish, who had helped us build EMILY's List in Texas, to join the staff and help build the EMILY's List Majority Council. With apologies to Sherry, the team initially became known as "the Judies." All three women were textbook exemplars of the market that formed the Majority Council. They were passionate about helping women, determined to see more women in office, and ready to work night and day. In addition, I hired Dee Ertukel as director of development and administration to expand our membership and upgrade members to the Majority Council level. Dee had been finance director for Dianne Feinstein's statewide races. Eventually, the team grew, and we changed its name from "the Judies" to the Majority Council Development Team or, as it became known, the Team.

As the Majority Council grew, we increasingly upped the ante. It wasn't easy. In other PACs, donors expected to get access to a high-ranking senator. For every donation, they expected votes in return. We were different. We were the antithesis of the PACs that represented, for example, huge corporate special interests, the tobacco industry, pharmaceuticals, and defense, each of which had legislation they wanted passed. We were mission driven. Our challenge was that women weren't used to funding politics and writing big checks, so we had to build a mission-driven female, large-donor funding base. Having a personal connection with like-minded Judies, explaining in detail what we were doing with their money, setting out "smart" strategies, and driving the emotions of our mission all helped build the large donor base that became the Majority Council.

At first, when I "added the zeros" to my requests, I was nervous. Was I asking for too much? Would I get turned down flat? And what if raising more money for EMILY's List diminished our ability to raise money for candidates? But I pinned my hopes on the idea that our members wanted our women to win. They had already seen how much difference their contributions made, and bigger donations to EMILY's List would improve the way their candidate contributions were spent and would lead to even more success in electing women.

Because we had been successful time and again, our generous and committed members were willing to trust us and write bigger checks. Eventually, the Team developed such strong relationships with the Majority Council members that they were asking for five-figure contributions, often without my even having to participate. And, like good political venture capitalists, we allocated our resources according to which campaigns needed our resources the most.

WE WERE CREATING OUR own full-service political operation. To that end, we took on as our new political director Mary Beth Cahill, a savvy political consultant from Boston who had worked for Rep. Robert Dornan (D-MA), Rep. Barney Frank (D-MA), Sen. Patrick Leahy (D-VT), and, most recently, Rep. Les AuCoin (D-OR) in his losing Senate campaign against Republican incumbent Bob Packwood. To help implement the program, Mary Beth also brought on two veterans of the AuCoin campaign, Joe Solmonese, a former aide to Michael Dukakis, as deputy political director, and Jeannie Duncan, a talented writer who worked with me on the newsletter and other marketing materials.

In August 1993, our new training director, Ellen Moran, launched our first Campaign Management Training School, with thirty-four prospective campaign managers from around the country coming to Washington to hone their political skills, paying particular attention to the dynamics of running a woman's campaign. Attendees included staff for ten congresswomen who had been endorsed by EMILY's List in 1992, as well as those working for 1994 senatorial and gubernatorial candidates.

"First, we trained people to be fund-raisers, since that was really our bailiwick," said Mary Beth. "Then, we began training them to be campaign managers, field operators, researchers, and press secretaries. We had to teach them how to spend the money well, how to put together good campaigns and really compete."

Ellen Moran developed a novel way to train new managers. A cornerstone of the training was building a campaign plan based on data from a campaign simulation in which fictitious candidate

Helen Winter, a pro-choice Democrat, runs against fictitious Wally Schroeder, a Republican and a former football coach, for an open congressional seat in the state of Delusion, which happens to be situated between the states of Ambivalence and Denial. During the day, Ellen would bring in consultants who would describe how their roles in campaigns worked. For example, campaign professionals such as pollster Celinda Lake taught trainees how to read and use a poll. Then, at night, the trainees used that day's newly learned expertise — whether it was how to target the voters they wanted to reach, what messages to send them, or how to raise the money to fund their plan — to build a campaign plan for their fictional candidate.

Before long, the trainees were completely immersed in developing winning strategies for this fictional candidate. Playing the role of Helen Winter, I let them know just how difficult some candidates could be. I refused to approve an adequate budget, make fundraising calls, or agree to what the campaign's message should be — just as candidates do in real life. Before the 1994 election cycle was out, no fewer than 125 fund-raisers, 59 campaign managers, and 40 press secretaries had enrolled in our training sessions. By the end of long days and nights, the teams were exhausted but had more than a rudimentary sense of how a campaign should operate.

MEANWHILE, HILLARY HAD BEEN appointed chair of the President's Task Force on National Health Care Reform — the most ambitious assignment ever undertaken by a First Lady. Those who really knew Hillary — and few did — should not have been surprised by the appointment. Ten years earlier, when Bill was governor and she was first lady of Arkansas, he had appointed Hillary chair of a committee to reform educational standards in the state. As *Politico* magazine reported, she was "open and more accessible. She was demanding, exacting and exhausting. And she was policy-first but politically astute, pragmatic, even calculating." "It was long and grueling, but there wasn't a person who felt like they hadn't been heard," said one participant. "Hillary let everybody say what they wanted to say."

Most importantly, when Hillary's proposals were enacted, Arkansas public schools went from being among the worst in the country—the state had more illiterates than college graduates—to having "the best standards we'd ever had in Arkansas," as the chairman of the state Board of Education put it.

But Washington wasn't Little Rock, Arkansas, and revamping America's multi-trillion-dollar health-care industry was a far bigger task. With tens of millions of people uninsured, it was also an essential one that involved taking on entrenched interests with hundreds of billions of dollars at stake. It was, as one journalist put it, like "scaling the Mount Everest of social policy."

On September 28, 1993, when Hillary testified about her health-care-reform proposal—all thousand-plus pages of it—before the House Ways and Means Committee, it was the first time a First Lady had testified as the key witness on a major piece of legislation. She used no notes. It was clear that she knew her stuff, and it was so compelling that it prompted Sen. Strom Thurmond (R-SC), who was notorious when it came to women, to jump to his feet when Hillary finished her presentation. "He said, 'I really agree with the little lady, I agree with her,'" recalled Hillary. "But all the other Republican senators were pulling him to sit down, and shut up." Later, Thurmond went up to Hillary and extended a handful of melted candy to her.

Even though she personally received high marks, the initial response to "Hillarycare" was not good. Liberals didn't like the way it was financed or the idea that each of the fifty states could go its own way. Conservatives hated the "big government" aspect. There was a lot at stake here. Health-care reform was the centerpiece of the Clinton administration's legislative agenda. If it passed, Clinton would easily win reelection in 1996. But there was no way the Republicans would let it pass, and the battle had just begun.

The Clintons had other problems as well. The *New York Times* had been investigating a failed real estate investment by the Clintons in the Whitewater Development Corporation in Arkansas, which took place when Bill Clinton was governor. Even though the Clintons lost money on the investment, David Hale, a former municipal judge and banker in Arkansas, accused Bill Clinton, then governor of Arkansas, of pressuring him (Hale) into making an illegal $300,000 loan to the developers of Whitewater, Jim and Susan McDougal.

What all the convoluted and murky allegations added up to was unclear, but, as with Vince Foster, that didn't matter. The media — both mainstream and otherwise — were on the warpath. The *New York Times* and *Washington Post* were relentless. Even though this was a pre–Fox News era, a new right-wing media landscape was ascendant. Funded by billionaire Richard Mellon Scaife, the *American Spectator,* a small conservative monthly, poured more than $2 million into "investigative reporting" asserting that the Clintons were behind Foster's death, probing Whitewater, and promoting allegations about Bill Clinton's sexual activities. Thanks to a new right-wing media food chain, the *Spectator* findings were often picked up by conservative radio host Rush Limbaugh. Rupert Murdoch's *New York Post* and the hard-right *Wall Street Journal* editorial pages gave the attacks — true or not — the aura of mainstream credibility.

All this meant that Hillary was taking her health-care-reform show on the road in the crosshairs of the right. Republicans, libertarians, movement conservatives, and the health-insurance lobby launched a massive campaign asserting that there was no health-care crisis in America and that Hillary's plan was bureaucratic big government at its worst. The health-insurance lobby put on a $20 million yearlong TV ad campaign featuring Harry and Louise, a forty-something suburban couple, bemoaning "big government" dictating what medicines they had to take.

The result was that when Hillary traveled the country to promote health-care reform, she was stunned to encounter the most threatening environment she had ever faced in her life. "I did rallies in

Seattle, which you think of as a very liberal place," Hillary told me. "Portland, Oregon, you think of the same way. But the Secret Service was taking weapons off dozens of men . . . Shock jocks on the radio were whipping people up to come teach me a lesson. It was really frightening. They [the Secret Service] were telling me, don't go out and speak."

HILLARY WENT AHEAD and spoke anyway, wearing a bullet-proof vest, but the political environment for the upcoming 1994 midterms couldn't have been worse. Having endorsed dozens of candidates in various races, EMILY's List was particularly focused on Ann Richards, Kathleen Brown, and Dawn Clark Netsch in gubernatorial races in Texas, California, and Illinois respectively, as well as on Dianne Feinstein's battle to win a full six-year term in the Senate.

One of the first tests of our new strategy was Dawn Clark Netsch, a former law professor who had become Illinois state comptroller and, initially at least, seemed like a hopeless long shot, having polled a paltry 14 percent of the Democratic vote leading into the 1994 primary. Our members contributed nearly $275,000 to Dawn's race. In addition, our new deputy political director, Joe Solmonese, moved out to Chicago for a month to oversee Dawn's fund-raising. We also sent political consultant Page Gardner there for two months to provide on-the-ground political support.

In addition, we had conducted our own focus groups, and the data suggested that voters liked the idea that women candidates meant a change from traditional backroom, cigar-smoking politicians, from politics as usual. Yet when they saw some of the traditional ads women candidates were running—with women sitting behind a desk signing documents, or touring a construction site in high heels and a hard hat—the focus groups pointed out that the women looked just like ordinary politicians. Voters didn't want that. The dressed-for-success look was what men did. As one participant said, "Nobody would really walk through a construction site in a nice suit, pearls, and high heels. Who are they kidding?" They wanted to see that women really were connected to their lives. We shared our data with media consultants and suggested they present women candidates to voters in ways that conveyed their more authentic, human aspects.

Saul Shorr, Dawn Clark Netsch's media consultant, certainly understood what we meant. He filmed an ad in which Dawn was shooting pool and made an incredible shot. The tagline was, "Dawn Clark Netsch — she's a straight shooter!" Dawn went up in the polls by more than 30 points and won the March gubernatorial primary with 45 percent of the vote in a three-way race.

One of the most important races in this 1994 cycle was for Dianne Feinstein's Senate seat in California. Dianne had first come to national prominence in 1978, when she announced the assassination of San Francisco mayor George Moscone and Board of Supervisors member Harvey Milk. As president of the Board of Supervisors, she served out the remainder of Moscone's mayoral term and was twice reelected in her own right. Over time, she cultivated an image of being tough but caring, and, after her narrow loss in the 1990 gubernatorial race, she had gone on to win her Senate seat two years later. There, she quickly established herself as an enormously effective senator in her role as the author of the 1994 Federal Assault Weapons Ban, overcoming the powerful National Rifle Association.

Dianne had won comfortably in 1992, but, because that was a special election to fill out the remainder of Pete Wilson's term, she was up for Senate reelection just two years later. This time, she faced Rep. Michael Huffington, a Republican congressman who had inherited a fortune from his father's oil business.

And who exactly was Michael Huffington? Among other things, he was the husband of Arianna Stassinopoulos Huffington, later of *Huffington Post* fame. At the time, she was known for her ties to the cult Movement for Spiritual Inner Awareness, led by a guru named

John-Roger, and was so ambitious she had been dubbed "the Sir Edmund Hillary of social climbers." As for the candidate himself, he had virtually no legislative record. He didn't talk to the press. He put forth so little substance that *The New Yorker* described him as "a tabula rasa" and "the purest candidate of all — almost untouched by experience." *Vanity Fair* compared him to Chauncey Gardiner, the hero of Jerzy Kosinski's novel *Being There,* who utters banal and meaningless platitudes that are received as profound cosmic truths.

Even Republicans saw Huffington as an empty suit. According to Hazel Richardson Blankenship, a Republican activist in Santa Barbara, Huffington's congressional district, "Huffington has never stood up for anything: I've never seen a position paper, a press conference, even an ad where he's said anything." "He's nonexistent," a fellow California Republican representative told the *Los Angeles Times.*

Nevertheless, Huffington did have money, and he had won his congressional seat after putting up $5.2 million of his personal fortune to finance the race. It was the highest amount ever spent on a congressional race, and cost him $41 per vote. Then, after he had served just four months in Congress, he announced he was prepared to do the same thing running for the United States Senate.

All of which made for an extremely bizarre race. A January 1994 poll showed Dianne Feinstein with a 53–25 percent lead over Huffington. But in February, ten full months before the election, Huffington began a massive media blitz. By the end of the month, it was clear that this was going to be perhaps the costliest Senate race ever. Dianne had spent $8 million to win in 1992, and Huffington was willing to spend nearly four times as much — $30 million, almost all of which was his own money.

Initially, Huffington's TV ads simply pictured him talking about personal values and the limited government he supported. But in May, he flooded the airwaves with one of the most misleading attack ads I'd ever seen. It was known as the "special-interest jukebox" ad

("Feinstein's a special-interest jukebox — put in your money and get what you want"), and it accused Dianne of doing special political favors for big companies in exchange for campaign contributions.

As revealed in the *San Francisco Chronicle,* however, the truth behind the ad's accusation was quite different. In fact, in August 1993, an executive at Raytheon Corporation, which employed more than one thousand people in Huffington's congressional district, asked Huffington to support a waiver to a trade agreement so Raytheon could sell a ship-defense system to Taiwan. Huffington refused to do anything, and that provided material for Feinstein to attack him as a do-nothing representative who was jeopardizing jobs in his district.

On the Senate side, Raytheon sought support for an amendment, sponsored by Sen. Frank Murkowski (R-AK), to open up defense trade with Taiwan. Feinstein supported the Murkowski amendment and wrote to Secretary of State Warren Christopher in its behalf. The amendment became law. It was true that Raytheon gave $2,000 to Dianne's campaign, but there was no reason for either her or Huffington to be against the bill. Raytheon was getting no "special break," because the bill helped not just Raytheon but its competitors as well. The bill saved hundreds of jobs in Huffington's district, and it did so simply by lifting government restrictions that had been put in place only to appease China. In truth, Huffington was serving China's interests, not his constituents'.

But the millions of Californians who saw Huffington's ads didn't know that. They now saw Dianne as a captive of corporate interests — as if $2,000 could buy her integrity. But with ads that totaled $4 million on the air in May, Huffington abruptly chopped 19 points off Dianne's lead in just one month. She was now leading only 48 to 41, the election was still six months away, and he was sitting on a huge bankroll.

I was horrified. The bottom line was that Dianne was no longer a sure bet, and if she lost, that potentially threatened the Democrats' hold on the Senate. "At that point, we realized this was going to be a negative slugfest and that Dianne's stellar record could be swept aside by Huffington's attacks," said Mary Beth. "Even the best coverage in the press can't stand up to negative ads on TV."

As it happened, we had recommended Feinstein early on and had raised more than half a million dollars for her. But with Huffington's vast fortune, it was now clear that wouldn't be enough. I sounded the alarm: "Feinstein needs help."

So, we sent Dee Ertukel, EMILY's List's director of development, back to California to help Feinstein. In addition, several Feinstein staffers attended our campaign-fund-raising boot camps.

Because 1994 was a nonpresidential election, we were concerned that many of the women voters who had been so excited in 1992 would stay home and not vote this time. Dianne needed every woman's vote she could muster, so we put together an effort to motivate key blocs of women.

That took the form of giving Ellen Moran a leave of absence so she could work with the California state Democratic Party on a project called WOMEN VOTE!, targeting women who voted only occasionally. "We recognized early that high turnout among women voters was going to be crucial in this race," said Mary Beth Cahill. "We worked with the Democratic Party in California on the get-out-the-vote strategy, finally sending Ellen Moran to manage the GOTV campaign."

Specifically, that meant focusing on about one million California women who had no history of voting in the last two nonpresidential elections. We also targeted newly registered women voters. We knew getting these women to vote would be vital to both the Feinstein campaign and Kathleen Brown's bid for the California governorship, as well as down-ballot races. So, Ellen Moran developed and implemented a plan to encourage women to vote using the vote-by-mail option that was rarely used by Democrats.

*   *   *

ON JUNE 24, HILLARY invited EMILY's List and our Majority Council members to a tea at the White House. After saying a few words to all of us, Hillary and I stood in a photo line greeting everyone who attended. I started to leave the photo line, because, frankly, I thought people would want their photo with Hillary, not me. But Hillary insisted I be there with her. Over the years, I often saw those pictures sitting prominently on bookshelves in the homes of various EMILY's List members. It was a big help in our fund-raising with the Majority Council.

And we needed all the help we could get. After months of wrangling over health reform, it had become clear that the centerpiece of the Clintons' initiatives was in deep trouble. In August, Senate majority leader George Mitchell introduced a compromise plan, but even it did not have enough votes to pass. The defeat left the Democrats severely wounded, with Hillary branded as a "big government liberal."

At the same time, the Republicans escalated their war on the White House. After roughly six months of investigation, special prosecutor Robert Fiske had released the preliminary finding in his fast-moving Whitewater investigation: first, he found no evidence that the Clinton White House or the Department of the Treasury had tried to influence the Resolution Trust Corporation investigation into Whitewater. He also concurred that Vince Foster's death was a suicide and had nothing to do with Whitewater. For a brief time, that looked like it would be the end of Whitewater.

But instead, because of a newly signed law, an ultraconservative "special division" panel of three federal judges was charged with picking a new special prosecutor. They chose forty-eight-year-old Kenneth Starr, a staunch conservative who had offered to write a friend-of-the-court brief for Paula Jones, who was suing the president for alleged sexual harassment. This was turning into another political witch hunt. Meanwhile, the House Republicans, having declared war on the administration, had a new field marshal to lead them out of the wilderness.

A man of many contradictions, fifty-one-year-old House minority whip Newt Gingrich of Georgia had briefly tried marijuana in the sixties, but ended up being repulsed by the counterculture and, seemingly, wanted to strike back, leading what the *New Republic* called "Revenge of the Squares" against an enemy he delighted in calling "the Great Society countercultural model." To that end, in the nineties, Gingrich pioneered the "the politics of personal destruction" that has since taken over the GOP. As he put it when he addressed a gathering of Young Republicans, "I think one of the great problems we have in the Republican Party is that we don't encourage you to be nasty."

That was not a problem Gingrich had. In a televised interview, his mother, thinking she was speaking off the record, told broadcaster Connie Chung that Newt referred to Hillary as "a bitch." His idea of "pro-family" legislation was to remove children from their families and put them in orphanages if they were born out of wedlock or to poor mothers.

But regardless of what one thought of him, Gingrich had a vision and a strategy that were able to pull together the Christian right and fiscal conservatives with a right-wing populist message that prefigured the Tea Party by more than a decade. On September 27, six weeks before the November elections, that message was presented loud and clear to the American people in the form of the Contract with America, written by Gingrich and Dick Armey (R-TX). An unprecedented document that specified policies the Republicans promised to enact in the areas of crime, fiscal responsibility, welfare, national security, job creation, and more if they won the midterms, Gingrich's Contract with America had the effect of transforming the midterms into a national referendum on Clinton at a time when health reform had just failed. Soon enough, we would learn how effective it would be.

★   ★   ★

A WEEK BEFORE the election, I flew to California to campaign with Dianne Feinstein and gubernatorial candidate Kathleen Brown. After spending the day flying south with Kathleen, I ended up in San Diego, ready for an early-morning event the next day with Rep. Lynn Schenk, one of our 1992 winners. When nobody could hear us, Lynn leaned over and told me that her nightly polls showed her dropping 5 points in the past two days. I flew back north with Dianne, stopping in two towns before ending up at a rally in San Francisco. On the way, we had a rare spot of good news: the press was reporting that Michael Huffington employed an undocumented immigrant as a nanny and did not pay employment taxes. The race with Dianne was tight, and, coming at a time when California was roiled by anti-immigrant fervor, that had to hurt him.

I had planned to fly out to Salt Lake City to have breakfast the next day with Rep. Karen Shepherd, a woman who seemingly had done the impossible by becoming a pro-choice Democratic member of Congress from Utah. But late that night, Karen called to cancel. Tracking polls showed her dropping 9 percent in the past week, and she had to record a last-minute TV spot for the final days.

All over the country, things were not going well. I put in a call to Texas to Ann Richards's fund-raising director, Jennifer Treat, because I was very worried about Ann's reelection race. When the race started, Ann had a 67 percent approval rating, was touted as having the potential to become the first woman president, and was going up against an opponent who had never won a race.

But she was facing George W. Bush, who, as the son of a former president, had fabulous name recognition and an aggressive campaign orchestrated by Karl Rove. According to Jennifer, the polls showed Ann and "Dubya," as Bush was known, neck and neck, with a huge undecided percentage — not a good sign for an incumbent. I was beginning to fear that the bottom was falling out.

I HATE ELECTION DAY. It's my least favorite day of the year. For one thing, unless I'm working to get out the vote, there's really nothing for me to do. I can't raise more money, there's nothing left to say to reporters, and much of the EMILY's List staff is scattered around the country helping the campaigns.

And the worst of it is the exit polls. Getting the results from the early exit polls is one of Washington's favorite games. Inevitably, someone gets those "tightly held" results, professes not to tell anyone, and the next thing you know the poll numbers are flying through D.C.

And they're often wrong. But even though I know the exit polls are often wrong and I tell myself every time that I'm not going to pay attention to them, I can't help myself. I jot down the numbers, I moan and groan, I yelp with excitement, and I pass the numbers on to my nearest and dearest.

Bad as many Election Days are, November 8, 1994, was the worst. It wasn't entirely unexpected. I had seen evidence of the free fall myself during my California swing. And on Monday, the day before the election, it was widely reported that the Republicans were likely to take over the Senate and might even win the House for the first time since the Eisenhower era.

As usual, I worked late and stayed on in the office with staff members, various friends, and steering committee members dropping by to follow the election results. We brought in food and opened a bottle of wine as we set up televisions and computers in the conference room where we had all our races posted on the wall. As we tracked the incoming results, we put a big *L* or *W* by each race. It didn't take long for me to realize that we had too many *L*'s that night.

Then, at 9:00, Dan Rather came out with it: "Bill Clinton's worst nightmare seems to be coming to pass: . . . [incoming Speaker of the House] Newt Gingrich, Senate majority leader Bob Dole, a real possibility Republicans will pick up forty seats net gain or more for control of the House for the first time in four decades, and may also get control of the United States Senate for the first time in eight years."

My worst fears were coming true for candidates like Lynn Schenk, Karen Shepherd, and scores of Democrats across the country. Karen went from a 9-point lead a week before the election to losing by almost 10 percent—a 19 percent drop in under ten days! There was nothing she or EMILY's List could have done.

All over the country, Democratic candidates were in free fall. Voters had wanted change in 1992 when they elected Bill Clinton and our EMILY's List women candidates. But nothing had changed. Health-care reform, the number-one objective of the Clinton administration, had been repudiated even though Democrats controlled the White House, the Senate, and the House. Voters were mad, and they were blaming Democrats.

As the pundits put it, the election shook Capitol Hill like an earthquake. Newt Gingrich called it a truly historic event. This was the Republican Revolution. In the House, the Republicans picked up an astounding fifty-four seats in the House and eight in the Senate to win both houses for the first time in forty years. At the time, it was the worst midterm defeat in modern history. A lot of the blame was placed on Hillary. The idea of her as a copresident was not going to fly. Moving into the future, no doubt, she would take on a more traditional, ceremonial role as First Lady.

In the final days, voters essentially decided to throw the majority party out of office. Their wholesale shift made it almost impossible for Democrats in marginally Democratic districts to hold on. We lost seven good congresswomen that night, and I was very sad. We had elected four new women to the House, but their victories occurred because they had won Democratic primaries in Democratic districts. We were unable to add any new congresswomen from competitive districts.

Among the gubernatorial races, Kathleen Brown lost in California, and Dawn Clark Netsch lost in Illinois. Worst of all, Ann Richards lost to Bush by 8 points. That was a true heartbreaker.

As for the Senate, the most important race to me was Dianne's battle against Michael Huffington. At 11:30 p.m. Eastern time, just thirty minutes after the polls closed on the West Coast, the race was still deemed too close to call. Having been outspent two to one in the most expensive senatorial race in U.S. history, Dianne was clinging to a small lead, but hundreds of thousands of absentee ballots had yet to be counted, and Huffington refused to concede. Traditionally, Republicans picked up votes from absentee ballots, but this time I hoped our women voters had used the vote-by-mail option and would boost Dianne. It was not until two days later that Dianne was officially declared the winner, by 165,000 votes.

Her victory was enormously important, of course, but I soon found out there was another story behind it that gave me even more hope for the future. Mary Beth and Ellen had worked hard on our WOMEN VOTE! campaign with the idea that it would be a huge help to Feinstein. Several long months later, when the data from the WOMEN VOTE! project finally came in, the numbers were so staggeringly high that we didn't believe them at first. Of the 902,575 women targeted by the campaign, 416,594 voted either by absentee ballot or at the polls, a turnout of 46 percent! That was an extraordinarily high turnout for ordinary voters in midterms, but for a pool of women who had *never* voted during midterm elections, it was astounding. According to the California Opinion Index, women gave Dianne a 13 percent advantage, whereas men preferred Huffington by 9 percent, creating a huge, 22-point gender gap. Clearly, these women provided the margin of victory for Dianne and other down-ballot candidates, including Rep. Jane Harman (D-CA), who won by only 812 votes.

Devastating as the 1994 midterms were for the Democratic Party, for us at EMILY's List they marked a new era in our development. We were no longer just a fund-raising operation. We had become a multipronged, full-service political operation. In addition, I was buoyed by the success of WOMEN VOTE! in California and saw that it had enormous possibilities if we deployed it throughout the entire country. I was very unhappy to see good women lose, but the next day, when I heard the incoming Speaker of the House, Newt Gingrich, expound upon the vast array of changes the Republicans would make, many of which would be devastating to women and their families, my feelings turned to anger. This was not over yet.

Thirteen

LEAPING AND CREEPING

SOCIAL CHANGE, I HAD LEARNED, was a matter of "leaping and creeping." At times, when conditions were right, it could take place in a massive wave, as it had during the Year of the Woman, in 1992. But more often, in the absence of a unique phenomenon such as Anita Hill's testimony, it took place incrementally, in small steps. In the wake of the Gingrich Revolution, we were now entering a "creeping" phase.

At times, "creeping" was a generous way of characterizing what was going on. When the One Hundred Fourth Congress was seated, in January 1995, the Democrats sat there shattered, as Gingrich launched one assault after another on everything we held dear. There would be draconian cuts in programs for low-income housing and child care. The elderly and the disabled could expect the same. A new health-care bill was out of the question. Meanwhile, the Republicans wanted to *lower* taxes for the rich.

Just a few days into that awful session, I was invited by House minority leader Dick Gephardt (D-MO) to meet in his office with him and Rep. Martin Frost (D-TX), the new chair of the Democratic Congressional Campaign Committee (DCCC). Gephardt had been newly installed as party leader, succeeding Tom Foley (D-WA), whose 1994 defeat was unprecedented for an incumbent Speaker of

the House in modern times. Gephardt's first order of business was
to rally the troops and to check the party's support systems.

I entered the minority leader's suite and took a seat facing Gephardt and Frost, who were sitting on the sofa. I couldn't tell whether
Gephardt was shell-shocked or simply bored, but in either case, he
was *not* engaged. For his part, Frost, a stocky, balding representative from the Dallas–Fort Worth area who now had oversight of the
party's congressional races, diligently took notes on a yellow legal
pad. Being in control of Congress was so much a part of the Democrats' culture that they didn't know where to start or how to rebuild.

Years earlier, I had learned that the only way to get the attention of the party establishment, the only way to break through the
glassy-eyed looks, was to talk numbers. So, I told them that EMILY's
List had raised almost $5 million for House candidates in 1994. This
money was not available to male Democrats, I explained, so each of
our twenty candidates that year had the advantage of support from
both the party establishment *and* substantial funds from EMILY's
List. If we wanted to take back the House, I argued, it was essential
to have more women candidates with the party's support.

Throughout the meeting, Martin Frost barely uttered a word, but
then, as I walked out, he came with me. There was something about
him that suggested he had just had a profound epiphany. "Ellen," he
said, "I've been crunching some numbers. Are you telling me you
raised an average of $200,000 for each House candidate?"

At the time, the average *winning* House race cost a little more
than $500,000. It had finally dawned on him that we had a lot to
offer.

"Yes, Martin. That's what I've been trying to tell you. If you recruit more women, you're going to have candidates with more financial and political support. So, work with us to help women, and
we can take back the House."

Bingo. He got it!

For roughly a decade, we had battled male incumbents within the party. But finally, they realized that this "stupid girls' organization with the silly name" was actually raising a lot of money, and that the more women voted, the more it would help not just women Democratic candidates but Democratic men as well. From that point on, we were valuable allies — partners, not some fringe organization.

It also meant that our role was changing from being merely a fund-raising operation to becoming a real political operation, one that would grow and grow. And that in turn meant that the credibility of women candidates would grow, and that the women we endorsed could move up the ladder within the party and play key roles on the national stage.

But first, now that Newt Gingrich had become Speaker, we had to fight back. With his cavalier disregard for women's interests and his relentless attacks on women's rights, Gingrich exemplified everything we disliked about right-wing Republicans, all in a profoundly unappealing package of arrogance and bravado.

Our new fund-raising appeal added a message that spoke to the rage and frustration of progressives and called for them to "Boot Newt!," to take back Congress and protect issues important to women and families. "Every day I read the morning paper and I get so mad, I can't wait to get in my car and drive to EMILY's List to get to work," the appeal began. It turned out to be the most successful opening line I ever wrote.

WITH MARY BETH CAHILL leading the way, we went national with the WOMEN VOTE! campaign. Conventional wisdom had it that the reason for the catastrophe in 1994 was that the Democrats had lost the vote of "angry white men." But, as Mary Beth pointed out, those were the so-called Reagan Democrats, who had already abandoned the party a decade earlier for Ronald Reagan. The conventional wisdom was dead wrong.

The real story, Mary Beth discovered, was that there was an enormous drop-off of *sixteen million* women voters between 1992 and

1994. A huge portion of those lost voters were women without college degrees, and their absence at the polls was devastating to the
Democrats. That meant that getting non-college-educated women
to the polls was the key to winning back the House and the Senate, and to getting Bill Clinton reelected. WOMEN VOTE! gave us
a clearly defined political strategy for victory. At our tenth anniversary luncheon, on May 1, 1995, President Clinton helped launch
WOMEN VOTE! as a national initiative.

For the first time, we were working hand in glove with the party
— and we were leading the way. We put together a $10 million mul-
tielection get-out-the-vote strategy targeted at women in key states
as a counterweight to the grassroots efforts of the Christian right.
Polls showed that women were alarmed by a wide range of Republican policies, including budget cuts to student loans, job training,
Medicare, and Medicaid.

Working with the party, we began buying and enhancing voter
files, overlaying demographic information from drivers' licenses
with files by age, income, education, and other variables. Then we
set about targeting these women voters with specific messages tai-
lored directly to their concerns, contacting them by phone and mail
to encourage them to vote.

We carefully scoped out which incumbent Republicans might be
vulnerable, and we found possible openings in California, Illinois,
Maine, Michigan, New Mexico, New York, Ohio, and Washington
State. Then, we recruited new women candidates for those openings, helped them to hire political consultants, trained their staff,
and guided them through the election process. We made them feel
they weren't alone in the fight.

In the wake of the Republican landslide, however, doing all of
this wasn't easy. The case of Darlene Hooley, a fabulously warm
woman who served on the Clackamas County Board of Commis-
sioners, near Portland, Oregon, was somewhat typical. When Karin
Johanson approached Darlene about running for Congress, Darlene

said no. A few weeks later, Karin asked her again, but still Hooley said no. Soon, such rejections became the new office joke. No one wanted to run for Congress when the Republicans were in control of it.

MEANWHILE, IN CONGRESS, Gingrich drove his agenda so forcefully that many people thought President Clinton seemed almost completely irrelevant. Then, on April 19, 1995, a massive truck bomb was detonated outside the Murrah Federal Building in Oklahoma City by Timothy McVeigh, a twenty-six-year-old militia sympathizer, and Terry Nichols, killing 149 people. In response, Bill Clinton came alive as a unifying, healing leader. "Here was a president who had been by many people deemed not to be strong, who suddenly was being viewed as both sensitive and strong," said speechwriter Don Baer. "At that moment, perhaps for the first moment, he inhabited the presidency."

The calculus of power between the House and the Clinton White House began to change. Throughout the rest of the year, the Republicans continued their assault, but Clinton fought back. The hard-right House voted repeatedly to approve one bill after another to prohibit federal funding for programs providing abortions in federal prisons, for family-planning programs, for city-run facilities in Washington, D.C., and more.

But Clinton was a steadfast defender of our right to choose and vetoed any bill that undermined Medicare, aid to children, or the safety net for the poor. "As long as they insist on plunging ahead with the budget that violates our values . . . I will fight it," he said. "I am fighting it today. I will fight it tomorrow. I will fight it next week and next month. I will fight it until we get a budget that is fair to all Americans."

BY THE END OF THE SUMMER, there was still no agreed-upon budget, and on September 30, the government ran out of operat-

ing funds. After that, it limped along on a temporary budget exten-
sion while negotiations continued between Congress and the White
House.

Then, about five weeks later, a historic tragedy took place that
had the inadvertent consequence of putting Gingrich's pettiness
on display for all to see. On November 4, Israeli prime minister
Yitzhak Rabin was assassinated, a catastrophe that dealt a devastating blow to efforts to reach a two-state solution between Israel and
the Palestinians. President Clinton, who had been a genuinely close
friend of Rabin's, immediately flew to the funeral in Israel, and he
invited Newt Gingrich and Senate majority leader Bob Dole (R-KS)
to accompany him on Air Force One.

Clinton elected to stay up front in the plane's presidential suite
with former presidents Jimmy Carter and George H. W. Bush on
the trip to Israel, as well as on the return flight. Gingrich was seated
in the rear passenger suite with Senator Dole, and, when the plane
finally landed at Andrews Air Force Base near Washington, was
asked to deplane by the rear door.

In Gingrich's eyes, this was a monumental snub, and on November 15, he told reporters how humiliated he was. "You've been on
the plane for twenty-five hours and nobody has talked to you and
they ask you to get off the plane by the back ramp . . . You just wonder, where is their sense of manners? Where is their sense of courtesy?" Ever the professor of history, Gingrich declared his seating
assignment one of the worst put-downs of the century.

To Gingrich, such a historic humiliation called for an equal and
appropriate response, so he sent President Clinton an even tougher
version of the budget resolution, knowing full well that Clinton
would veto it. Later that afternoon, however, the White House released a photograph taken of Gingrich on the flight, talking with
the president and Majority Leader Dole. The photo, when juxta-
posed with his remarks, was devastating to Gingrich. The *New York
Daily News* marked the occasion with a cover cartoon of Newt as a
diaper-clad infant carrying a baby bottle. CRYBABY, read the head-

line. NEWT'S TANTRUM: HE CLOSED DOWN THE GOVERNMENT BE-
CAUSE CLINTON MADE HIM SIT AT BACK OF PLANE.

AND SO BEGAN a historic government shutdown. For six days,
eight hundred thousand federal workers were furloughed. The Federal Housing Administration couldn't process home sales. About
six hundred thousand elderly who depended on the Meals on
Wheels program were now at risk. National parks were closed. It
wasn't pretty. But Bill Clinton had stood his ground, and in the eyes
of most Americans, Newt Gingrich and the Republicans were responsible for shutting down the government.

Gingrich had overplayed his hand so much that before long,
Senator Dole, the likely Republican presidential nominee for 1996,
sensing that the government shutdown would hurt the Republicans, put together the necessary votes to approve the Clinton budget. Clinton had won, and suddenly the Democrats were back in
business.

And so was EMILY's List. Karin Johanson called Darlene Hooley
in Oregon once again, and this time Hooley wanted to run. In
Michigan, Debbie Stabenow, a longtime friend of EMILY's List, had
worked her way up the state senate and, after suffering a narrow
loss in Michigan's 1994 gubernatorial primary, emerged as a promising candidate for a congressional seat. She was a real natural and,
we felt, a good choice to challenge incumbent representative Dick
Chrysler, who was vulnerable, given his close ties to the Gingrich
Revolution.

In California, we needed someone to take on Rep. William
Baker, a Republican who was much too conservative for his district,
outside San Francisco. Judi Kanter suggested EMILY's List member
Ellen Tauscher, a business-savvy investment banker who was one
of the first women to become a member of the New York Stock
Exchange and who had been a major fund-raiser for Dianne Feinstein's successful senatorial campaign.

In New York, we recruited Carolyn McCarthy, a former Republican who had launched a gun-control crusade after her husband, Dennis, was killed and her son, Kevin, wounded when a gunman on the Long Island Railroad opened fire on passengers. She stood a good chance against Rep. Daniel Frisa, a Gingrich ally and a friend of the gun lobby, so we helped her hire consultants, taught her how to debate — the whole nine yards.

And in Indiana, we found Julia Carson, a local official who had little experience with modern sophisticated campaigns but had a strong grassroots base and a spectacular track record, having turned a $20 million debt in the local welfare program into a $6 million surplus. We encouraged her to run for Congress when her representative retired, in 1996.

IN APRIL 1996, we kicked off our WOMEN VOTE! campaign with a sold-out luncheon featuring Hillary Clinton as our speaker. "Women who never thought of themselves as political before are now finding their voices and understanding the importance of their votes," she said. Soon afterward, Democratic National Committee chairman Don Fowler joined me in announcing that the WOMEN VOTE! campaign would target Iowa, Michigan, Pennsylvania, Washington, and New Jersey, with seven or eight more states to follow.

By this time, having Bill and Hillary Clinton and the chair of the DNC in our corner, it was fair to say we were no longer outsiders. More closely allied with the party establishment than ever, we began to put our strategies to work.

At the time, no one had paid real attention to women voters as a monolithic bloc. The concept of the "gender gap" was not commonplace, so we set out to define the difference between men and women voters in terms of priorities and to show how that difference was important to the Democrats. To that end, in the spring of 1996, we commissioned a series of eight polls tracking women's preferences on issues. We released these "women's monitor" polls at press conferences to drum up interest in the gender gap. Sure enough, the position of women began to be discussed as a determining factor in the next election.

Focusing on non-college-educated women, we found a huge population living on the edge of economic disaster: wives who worked to support their families and came home completely exhausted just in time to cook dinner; women who couldn't afford to get their cars fixed when they broke, meaning they couldn't get to their jobs; women who were fed up with voting for politicians who said they were going to help but just ended up partying with lobbyists. No wonder many didn't even bother to vote!

Our campaign showed these women that voting could actually make a difference for them and their families. We said, if you want your child to be able to go to college, you should know that Mary Smith supports student-loan money, but Bill Jones, the Republican, is against it.

As the 1996 election approached, "gender gap" became the buzzword of the day. EMILY's List became the go-to place for reporters seeking information about it. Increasingly, data supported the rationale behind our campaign. A study conducted by the *Washington Post,* Harvard University, and the Kaiser Family Foundation showed that two-thirds of disaffected voters were women — most of whom were against cutting programs for the poor or the elderly, and were deeply concerned about the growing gap between the wealthy and the poor. And, of course, they were precisely the women we were targeting with our WOMEN VOTE! campaign.

SOON, WE SAW THAT the patterns of 1994 were reversing themselves. Republican women were demoralized and were considering not voting at all. Meanwhile, Democratic women in union households, minority women, and single women were energized as never before. Ultimately, we saw a 17-point gender gap in congressional matchups between Republicans and Democrats. Similarly, women gave President Clinton a 22-point lead over Sen. Bob Dole (R-KS), the Republican nominee. The data showed what we had been saying all along: the Democrats would win if the women turned out to vote.

When the election returns came in on November 5, 1996, the results were stunning compared to two years earlier. In one of the great comebacks in American political history, Bill Clinton became the first Democrat to win a second term since Franklin Roosevelt, beating Bob Dole by more than eight million popular votes and winning more than 70 percent of the electoral vote. Exit polls showed the merit of our strategy, with women supporting Clinton over Dole by 17 points.

Although we didn't reclaim the House for the Democrats, we narrowed the GOP's edge by five seats, and the women we endorsed fared well. Louisiana's Mary Landrieu became our sixth Democratic woman in the Senate. Jeanne Shaheen became the first woman governor of New Hampshire. And we elected no fewer than nine new women in the House — the second-biggest increase in history — including wins by Carolyn McCarthy, Julia Carson, Debbie Stabenow, Darlene Hooley, and Ellen Tauscher.

ON JANUARY 20, 1997, when Clinton took the oath of office for his second term, I was vice chair of the inauguration and sat on the Capitol steps with members of Congress to watch the swearing-in. Within the Democratic Party, WOMEN VOTE! was seen as having played an enormous role in Clinton's victory. We had won a new level of respect from the Democratic political establishment.

And yet, these relatively enlightened attitudes did not yet prevail in the party throughout the entire country. One place where that was particularly noticeable was Colorado, where we had two strong candidates in Dottie Lamm and Lt. Gov. Gail Schoettler, running for the United States Senate and the governorship, respectively, in the upcoming 1998 election. Both women were the front-runners in the Democratic primary but faced difficult general elections.

To help Dottie Lamm, Ellen Moran, EMILY's List's former training director, who went on to run the California WOMEN VOTE! campaign in 1994, went out to Colorado as Lamm's campaign manager and, by all accounts, was doing a terrific job raising money and getting the campaign in order. But, in the spring of 1998, I started to see stories in the press that both Dottie's and Gail's opponents were gaining ground. Before long, I went to Denver for an EMILY's List event and asked Ellen Moran about it.

"You got me," she said. "All our analysis and data show us way ahead in the primaries, and I have no idea where the stories are coming from. They just have to be wrong."

When I looked into it, I discovered that the stories came from members of the state political establishment, including a labor leader, the executive director of the state democratic party, and a political reporter. I could just imagine the scenario: the labor leader probably called the party director and said, "Our guy is gaining ground." The party director, excited by his inside knowledge, called the reporter and said he heard the guy was really on the move. And the reporter wrote it up. That's how the conventional wisdom was forged.

But when the voting was over, Ellen Moran's data was absolutely correct. The results weren't even close. Dottie won her primary with 58 percent of the vote, and Gail won with 55 percent of the vote. I guess the guys in the small network of "experts" just got carried away by their emotions.

OUR COLORADO CANDIDATES weren't the only women doing battle with the old boys' network. Tammy Baldwin was a popular member of the Wisconsin State Assembly, who clearly had potential for higher office. One of very few openly gay politicians at the time, Tammy called herself "a proud progressive," and she had been elected to the state assembly three times by wide margins from the liberal college town of Madison. Rumor had it that Scott Klug, the incumbent Republican representative in her district, was about to retire. Tammy was a natural to run for his seat — and she wanted to make sure she was well prepared.

So, in 1997, Tammy came to Washington, where she met with me and Karin Johanson. Less than 48 hours later, Klug announced he would not seek reelection, thereby cementing Tammy's decision to run. We assigned Jonathan Parker, one of our political advisers, to assess Tammy as a candidate and to assist her campaign, and we introduced her to her eventual campaign manager. We helped Tammy set up her money, polling, and media operations. Similarly, Tammy sent some of her local staff to Washington to be trained by us. All the pieces of our political program were coming together.

In November 1997, Tammy was one of more than a dozen candidates who came to our Washington media-training sessions, where top media consultants subjected prospective candidates to on-camera interviews that were recorded so that the candidates could be reviewed and evaluated.

Even though we thought we had finally made peace with the old boys' network, Tammy had one major problem. The national Democratic establishment didn't believe she had a chance. Rep. Steny Hoyer (D-MD), a rising star in the Democratic leadership, was in charge of recruiting congressional candidates for the party, yet he publicly backed a rival of Tammy's, Dane County executive Rick Phelps, whom he had known for years, in the Democratic primary. Phelps's staff included Jim Messina, an up-and-coming political consultant who had learned his craft at the EMILY's List campaign managers' school and was now working against us.

Karin Johanson, who was now working flat out for Tammy on our behalf, had been Steny Hoyer's chief of staff. "She was furious at Steny," said Tammy. "She went ballistic. She said, 'What the heck are you doing?' Karin just raked him over the coals." Meanwhile, EMILY's List members contributed more than $70,000 to Tammy during the early stages of the primary, not to mention all the other resources we were steering her way. It paid off handsomely, too. In September, Tammy eked by Rick Phelps in the Democratic primary by 1,514 votes.

Nor was Tammy the only beneficiary of our labors. By now, we had trained more than 250 staffers who were willing to work with pro-choice Democratic women candidates, and more than half were already working on candidacies we had endorsed. In addition, our seminars for the candidates themselves ensured that we were building relationships with campaigns and becoming familiar with their needs so that we could help them down the road.

Looking forward to November, Barbara Mikulski, Barbara Boxer, Carol Moseley Braun, and Patty Murray all had serious reelection challenges, the last three having been targeted by the Republican Senatorial Campaign Committee. All three were now running in a midterm year in which turnout would be much lower and would be vulnerable if women voters stayed home.

TO MAKE MATTERS WORSE, President Clinton had confessed to his affair with White House intern Monica Lewinsky. I was furious with him for lying about it to the country and putting all of his allies in a difficult position. For months, special prosecutor Kenneth Starr had been leaking salacious tidbits about the Lewinsky affair to the press, and when he released his final report, on September 11, the public devoured the lurid details. I didn't feel the affair was sufficiently serious to call for impeachment, but, as president of EMILY's List, I called on Congress to censure President Clinton. Then, I hoped we could put this tawdry mess behind us and move on.

Angry as I was with the president, I was even more furious at the Republicans, who were attempting to bring him down. In the House, the prospect of impeachment was looming. The scandal had motivated Republicans so strongly that they had amassed nearly twice as much money as the Democrats for the midterms. It had also provided enormous fodder for the Christian right, leading the

Christian Coalition to launch ferocious get-out-the-vote efforts and to distribute forty-five million supposedly nonpartisan voting guides.

Unfortunately, EMILY's List was trapped in the larger political environment and we had no choice but to move forward, do our research, raise money, send out mailings, work with our candidates, and make sure that as many women as possible won.

At Labor Day 1998, the situation seemed dire indeed, but we went into overdrive. Over the next two months, EMILY's List members came through as never before, contributing an additional $4 million before the election. Two other factors were crucial. For one thing, the Republicans' refusal to let go of the Lewinsky scandal seemed to have backfired. In addition, by targeting seniors, African Americans, Latinas, professional women, and women without college degrees with massive mailings in twenty-two states, we were able to turn the gender gap into votes for Democrats all over the country.

In the end, Carol Moseley Braun could not overcome the wealth of her Republican opponent, Peter Fitzgerald, who beat her by 3 points. But other than that, it was a terrific year. Barbara Boxer ended up being the first candidate for whom EMILY's List had raised $1 million. She spent it wisely, pulling away from her Republican challenger at the end. It was no surprise that Barbara Mikulski won easily, but so did Patty Murray, and we picked up another senatorial seat with Blanche Lincoln's victory in Arkansas. Jeanne Shaheen easily won reelection as governor of New Hampshire. In the House, every single pro-choice Democratic woman was reelected, and we gained seven new seats for women, including Tammy Baldwin's in Wisconsin. It was the biggest increase of Democratic women in the House in a nonpresidential election year.

As for the House of Representatives, the Republicans ended up having the worst midterm election results in sixty-four years for a party not holding the presidency. Although they narrowly retained the majority, the GOP lost five seats to the Democrats, at a time when it was widely expected that they might pick up thirty. EMILY's List could take some credit for that. According to our postelection analysis, in the states targeted by WOMEN VOTE!, fully 73 percent of our candidates won.

The results were enormously embarrassing for Gingrich in an election in which everyone expected a Republican sweep. By Friday, just three days after the election, Gingrich announced that he was stepping down as Speaker of the House and resigning from Congress in January. Newt had indeed been booted.

OUR NEXT TASK WAS to help the Democrats win back control of the Senate, where the Republicans had a 55–45 majority prior to the 2000 elections. Several of our women candidates were crucial.

One of them was already exceedingly well known. Just after the 1998 election, Sen. Daniel Patrick Moynihan (D-NY) announced he was not going to run for reelection in 2000, and Rep. Charlie Rangel (D-NY) immediately called Hillary Clinton, asking her to run for Moynihan's open seat.

"I thought it was absurd," Hillary told me. "I was flattered. And they may have liked me, but they had a very clear political self-interest in why this was good for them." With Rudy Giuliani as the likely Republican candidate, the Democrats needed a contender who could compete in terms of money and name recognition.

"I just kept saying, 'No, no, no, I'm not going to do this,'" Hillary later remembered. "But every week there'd be another story they would plant about how I'm thinking about it and they're talking to me and I'm meeting with them."

Initially, Hillary told the press, she wasn't running. But in January 1999, Hillary turned on *Meet the Press* one Sunday to see Bob Torricelli, then head of the Democratic Senatorial Campaign Committee, saying, to her astonishment, that he fully expected Hillary to run. Soon, a flood of New York friends who had supported Bill asked to talk to Hillary.

"I couldn't say no to all of them," she recalled.

Once they got their feet in the door, they began saying how much money they could raise, and whom they could get to endorse her. But Hillary still wasn't convinced.

Finally, in March 1999, Hillary came to New York, because HBO was putting together a special on women athletes called *Dare to Compete,* hosted by tennis great Billie Jean King, which was being shot at the Lab School, a public high school in the Chelsea neighborhood of Manhattan.

Sofia Totti, the captain of the girls' basketball team, introduced Hillary, and as Hillary came up onstage and shook her hand, Totti leaned forward and whispered in Hillary's ear. "Dare to compete, Mrs. Clinton," she said. "Dare to compete."

"That was the most telling argument that had been made to me," Hillary said. "I'd gone around telling all these young women to get out there, fulfill their destinies. And that's when I started thinking seriously about running for the Senate. I don't know that I ever would have crossed over without that young woman. She really called my dedication, my identity, and my commitment into question."

AT THE PRESIDENTIAL LEVEL, of course, the 2000 electoral season was dominated by the historic race between Vice President Al Gore and Texas governor George W. Bush. With the GOP party establishment plowing enormous resources into the elections — the National Republican Congressional Committee raised $36 million, and the Republican National Committee put in $100 million for voter turnout — and the Christian right going full tilt for Bush, WOMEN VOTE! was more crucial than ever before. We knew we had to get out the vote if the Democrats were to win, so in seven states we contacted eight million women, about double what we had reached in 1998. The races in which EMILY's List had candidates were particularly crucial to the Democratic Party's hopes of retaking the Senate.

When it came to Hillary's senatorial campaign, we raised plenty of money, but, given the Clintons' abundant electoral skills and resources, our help was far more crucial to other candidates. One of my favorite races took place in Michigan, where Debbie Stabenow was challenging incumbent Republican senator Spencer Abraham, who had entered the Senate in the Gingrich Revolution of 1994.

Debbie, you may recall, had hosted parties for EMILY's List at her home in Lansing, Michigan, when she was a state representative, and had worked her way up the ladder to the state senate. In 1994, she had campaigned for the Democratic gubernatorial nomination of Michigan. She lost narrowly but earned an enormous amount of goodwill within the party. In 1996, Debbie was elected to Congress, but after four years she was less than enthralled by the combination of non-stop fund-raising for her biennial campaigns and the lack of power that comes with being in the minority party.

"I made a personal decision that I wanted to be someplace where I could actually get things done," said Debbie. "Where I could work on policy." As a result, she decided to run for the United States Senate.

Given that a woman had never unseated an incumbent senator, Debbie's candidacy was a long shot from the start. In light of the difficulty of that task, Debbie's first step was to make sure she didn't have to go through a costly primary clash that would deplete her resources. Her most likely Democratic rival was Jim Blanchard, the former governor of Michigan.

"Blanchard had said if he was going to take on a Republican incumbent, he didn't want to have a primary [against another Democratic candidate]," said Debbie. "So, we created a strategy to make it very clear that he was going to have a primary."

At EMILY's List, we divided up the calls and announced that we were going to do everything we could for Debbie. By the following week, Blanchard was convinced he couldn't avoid a rough primary battle with Debbie, but he wasn't quite ready to back out.

"After all those calls, he phoned me and said, 'I want to meet,'" Debbie told me. "And he came over to my house, and I remember telling him that I really wanted to run, and EMILY's List would support me and I'd have a really good chance to win. He was quite surprised that I didn't defer to him."

And because she didn't, Blanchard dropped out.

BLANCHARD'S WITHDRAWAL MEANT that Debbie could challenge Abraham without the bother of a primary, but she still had another problem: money. This was an era before the Supreme Court's *Citizens United* decision allowed the Koch brothers and other billionaires to pour virtually unlimited funds into super PACs, but Spencer Abraham nevertheless was able to get enormous help from outside sources. Antitax activist Grover Norquist and his Americans for Tax Reform started placing pro-Abraham ads worth $1 million on the air more than a year before the election. Senate Republicans pressured high-tech companies such as Intel to raise money in support of Abraham's candidacy. By April 2000, Abraham had raised $6.4 million for his reelection campaign, roughly twice what Debbie had.

Americans for Job Security, a pro-Abraham, Virginia-based business group, placed attack ads in the *Michigan Chronicle* and the *Michigan FrontPage,* which had largely black readerships, hoping to drive a wedge between Debbie and her African American supporters. Right to Life of Michigan also went after Debbie. The message was clear: Republicans knew Abraham was vulnerable and were pulling out all the stops to save his seat in what was one of the key battleground states in the country.

At the time, it was unheard of for any political group to spend so much money on TV advertising so long before the election. "First, one group went on the air thanking him for something, and then another group, and another," Debbie said. "They had somebody on the air from the end of March till Election Day."

Yet, even with Abraham's huge cash advantage, Debbie and he were still in a dead heat in June. Knowing that Debbie would be massively outspent, EMILY's List sent fund-raiser Yael Ouzillou to Lansing to assess her campaign's finance operation. Our deputy political director kept in touch with the campaign on a regular basis to provide political advice, and another former EMILY's List staffer joined Stabenow's staff as research director. Before the election was over, no fewer than twelve thousand of our members contributed $1.4 million to Stabenow.

But the campaign became ugly. Among the first signs was the launching of Libberaldebbie.com, an attack site sponsored by Abraham. Business groups, many funded by pharmaceutical and insurance interests opposed to Debbie's health-care-reform proposals, spent nearly $10 million attacking her on radio and TV.

As the summer wore on, Abraham's relentless attacks took their toll. By mid-July, Abraham had taken a 3-point lead, 45–42. By the end of the month, he had an 11-point lead. By September 20, his lead had widened to 44–32. Soon afterward, a poll by the *Detroit News* showed Debbie trailing by 17 points, with time running out.

Typical for Debbie, when she heard about this disastrous poll, rather than get depressed, she went out and did five more events that day. At the same time, at EMILY's List, we hadn't given up on her either. To me, the fact that Spencer Abraham couldn't get above 50 percent meant that his attack ads were driving down Debbie's numbers but that voters still didn't like him. He and his allies had spent millions of dollars praising him to the heavens, but voters weren't convinced. I knew Debbie was a sensational candidate — warm, personable, telegenic, and an enormously likable campaigner — and I was certain voters would see how fabulous she was if only she had enough money to communicate with them.

Over the summer, Debbie worked tirelessly to raise money, but it was challenging. When she went to the Democratic Senatorial Campaign Committee fund-raisers for help, she didn't get very far. "In the end they didn't walk away from me," Debbie said. "But they would say, 'We think you're terrific, but first, we're going to help Harry, and then Fred, and Joe, and when we're done with those guys, we'll get to you.'" She was at the bottom of their list.

So, I put Debbie at the top of ours. I grabbed the staff and said,

"Look under sofa cushions, if you have to! Do everything you can think of to see how much money we can put together for Debbie Stabenow."

By the end of the day, Joe Solmonese, our executive director, came back to me and said he thought we could raise another $500,000 for Debbie. We started to work executing his plan. We organized special mailings and phone banks. We went to the Majority Council. We went to all our major donors to fund a special WOMEN VOTE! drive in Michigan. We were as motivated as we'd ever been.

I called Debbie to tell her the good news: EMILY was coming to the rescue. When I finally reached her on her cell phone, Debbie was caught up in the mundane minutiae of daily life, shopping at Nordstrom for a new blouse for a campaign appearance that night.

"I need to talk to you," I said. Debbie found a stairwell for privacy. "I know the polls," I continued. "But I have incredible news. The cavalry is on its way! I totally, absolutely believe you can win this. So, I brought the staff together, and I want this to be the top priority for the election, and we are going to raise $500,000 for you before the end of Election Day. And we are going to do a big WOMEN VOTE! campaign for you as well. Don't give up for one instant. You are fabulous, and we can do this."

"That's terrific!" she said. Her one and only televised debate with Abraham was coming up. "Just what I need going into this debate."

And it was. Abraham entered the debate overconfident. "He totally underestimated me," said Debbie. "He was obviously not prepared, and we had prepared for this as if it were the Olympics."

In addition, Debbie had put together a "rapid response" team to counter any inaccuracies coming forth from Abraham's camp. They called it Michiganfactcheck.com, and the media started using the team's information in all their news reports.

Suddenly, Abraham's lead vanished. By October 26, a poll showed Debbie and Abraham in a dead heat, and now she was matching him ad for ad. Then, a November 2 poll, just before the election, showed Debbie actually in the lead for the first time, thanks largely to likely women voters, who favored her by a 14-point margin.

OF COURSE, DEBBIE WASN'T our only senatorial candidate that cycle. In Washington State, Maria Cantwell was taking on a Republican incumbent, Slade Gorton. A former congresswoman who was defeated in 1994, Maria briefly gave up politics and went on to make a fortune as a tech executive at Real Networks. Now, she was back in the hunt, locked in a race with Gorton that was going right down to the wire.

Meanwhile, in New York, Hillary's presumptive Republican opponent, New York mayor Rudy Giuliani, had been expected to give her a run for her money. But in the spring of 2000, Giuliani found himself beset by a host of marital, political, and medical woes. When an extramarital relationship sparked a media circus, Giuliani announced his separation from his wife, and then discovered he needed treatment for prostate cancer. As a result, he withdrew from the race, and Hillary faced a less challenging battle against his replacement, Rep. Rick Lazio.

Another important race for us involved Mel Carnahan, a progressive governor from Missouri, who was running for the Senate against Republican John Ashcroft. Of course, Mel wasn't eligible for endorsement from EMILY's List, but he was staunchly pro-choice, and he had called me several times over the years asking for my personal financial support to fight anti-choice measures that the Republicans in Missouri were foisting on him as governor. When he decided to run for the Senate, we shared our polling data on women voters with him, and we were pulling for him in a tough race against Ashcroft.

But then, on October 16, Mel, one of his sons, and one of his advisers were killed when their small plane crashed on the way to a fund-raiser for his Senate campaign. With less than a month to go before the election, it was too late to take Carnahan's name off the ballot. But a week later, Missouri governor Roger Wilson, a Democrat, announced that he would appoint Carnahan's widow, Jean Carnahan, to the Senate if her late husband got more votes than Ashcroft. Suddenly, we had the possibility of putting yet another woman in the Senate. We immediately began raising money for Missouri WOMEN VOTE!, ultimately adding $150,000 to bring women to the polls.

ELECTION DAY, TUESDAY, November 9, 2000, is widely remembered by most Americans as the night Al Gore and George W. Bush fought to a draw, leaving the fate of the presidency in the hands of Florida, followed by bitterly fought controversies over hanging chads and butterfly ballots, inconclusive recounts, and numerous lawyers from both parties battling over the presidency. For those of us at EMILY's List, it was a memorable night as well, but the outcome was considerably happier for us than that of *Bush v. Gore.*

First, Hillary won in a breeze, beating Rick Lazio by 12 points. Then, in Missouri, which had been a tight race, Mel Carnahan was elected posthumously, beating John Ashcroft by 2 points. That meant that his widow, Jean, would take his seat in the Senate, and that we had added at least two women to the Senate. Given that we had also helped elect Jeanne Shaheen as governor of New Hampshire and Ruth Ann Minner as governor of Delaware, not to mention forty-one representatives, it was not a bad year.

But it wasn't over yet. The eyes of the nation were closely focused on the historic presidential spectacle unfolding in Florida, but we had more than our share of drama in both Michigan and Washington State. When it came to Debbie Stabenow's race against Spencer Abraham in Michigan, even on Wednesday, the day after the election, ABC pronounced the race too close to call. Not until Thursday was it reported that Debbie won by some 67,000 votes — out of 4 million. It was an incredible victory, the upset of an incumbent who, between him and his special-interest friends, had outspent her roughly three to one.

That meant we had added three senators: Jean Carnahan in Missouri, Hillary Clinton in New York, and Debbie Stabenow in Michigan. But Maria Cantwell's race in Washington was another cliff-hanger going into overtime. By Wednesday, with more than 1.6 million votes counted, Slade Gorton led by 3,000 votes, and both he and Maria had declared victory. In truth, the outcome was in the hands of half a million absentee voters, whose votes would not be counted for days.

I took considerable comfort in the fact that WOMEN VOTE! had put in lots of work on absentee voting in Washington. Because the secretary of state's office would post daily county-by-county returns, I would go online every day, determine what votes hadn't been counted, and crunch the numbers. I couldn't be certain, but it looked to me as if Maria was going to win by about 2,000 votes. A week later, Gorton was still leading, but a number of counties that leaned Democratic had not yet fully reported.

At the time, the entire country was still on tenterhooks over the Bush-Gore battle. It was unprecedented for the fate of the presidency to hang in the balance, as it did for more than a month after the election. The Cantwell-Gorton race in Washington was not quite as historic, but it was dramatic nonetheless, and the stakes were high: a Cantwell victory would leave the Senate deadlocked at 50–50.

In the context of so many cliff-hangers, it was satisfying to know that EMILY's List had raised more than $9 million for candidates, with a robust fund-raising presence on the Internet. We had also spent a record $10.1 million on WOMEN VOTE!

Two weeks after the election, Maria moved ahead to take a 1,953-vote lead, a razor-thin margin of 0.08 percent. There was a mandatory recount, but in the end, she and Debbie Stabenow became the first two women to defeat incumbent senators. That meant four new women senators — raising the total from nine women in the Senate to thirteen, and matching our results in the Year of the Woman.

It also meant that the Democrats and Republicans were tied at 50–50 in the Senate. With the Republicans in the White House, Vice President Dick Cheney was the presiding officer of the Senate, meaning that the Republicans remained in control.

But that was only temporary.

Fourteen

THE NEXT PIECE OF THE PUZZLE

IN EARLY SPRING OF 2001, at a staff meeting to plan our big Majority Council conference in June, I made a suggestion that went over like a lead balloon. I proposed that we invite as our main speaker at the conference Senate minority leader Tom Daschle (D-SD). The reaction was unanimous: bo-r-r-r-i-n-g. That response was not terribly surprising. Daschle had a reputation as safe, solid, bland, and colorless as white bread.

But there was a reason behind my suggestion. I realized, of course, that if the Democrats won just one more seat in the Senate, they would take control. Then, Daschle would be Senate majority leader. I had met Daschle several times and thought he was as a strong, if somewhat low-key, leader. We issued the invitation.

As it happened, I didn't have to wait until the next election for a power shift. Jim Jeffords, the Republican senator from Vermont and chairman of the Senate Education Committee, got so fed up battling with his own party that he became an independent and decided on June 6 to caucus with the Democrats. The bottom line: the Democrats now had control of the Senate. As a result, Democrats were taking over the chairs of each and every senatorial committee, and Tom Daschle became majority leader. Two days later he addressed the Majority Council conference.

Suddenly, I looked like a genius. When the new Senate majority leader walked into the room, we all leaped to our feet, cheering and waving napkins. He was a rock star. We had finally won back the Senate; the Republicans were no longer in charge. "The reason I am here today as Senate majority leader can be said in two words," he told us. "EMILY's List." At the close of his remarks, he apologized for having to leave. "I promised Jim Jeffords I'd mow his lawn."

WE HAD BEEN LOOKING forward to 2002 in large part because it was one of those magical "2" years. In 1992, you may recall, the Year of the Woman had been incredibly successful, thanks to two key factors. One of them, the fact that Anita Hill had unleashed an unprecedented wave of pro-woman political sentiment, was impossible to replicate.

But the other factor — redistricting — was an event that took place after the census every ten years. Every state with more than one House district redrew the lines, sometimes resulting in significant shifts in the partisan makeup of the district. All these changes created more open seats and vulnerable incumbents. For EMILY's List, this meant new opportunities to break through the gridlock and elect more women. The bottom line was that there were fifty-three open seats in the House — not as many as in 1992, but still nearly double the number that opened up in normal years. All of which meant that now, in 2002, we were looking forward to another big year.

We were in for a crushing disappointment.

One reason was that the entire political landscape was in flux. Indeed, just as we had benefited from entirely unforeseeable serendipity ten years earlier, in 2002 it could be said that the Republicans also benefited from an unforeseeable event. The horrifying attacks on the World Trade Center and the Pentagon on September 11, 2001, had dramatically changed the political discourse in the United States. National security became paramount. All of which cast a patriotic halo on the Bush-Cheney administration, on hundreds of Republican candidates across the country, and on George

W. Bush, transforming him into a wartime president and, for a brief time, the most popular president in the history of the United States, with a 92 percent approval rating. After 9/11, like all Americans, I found myself cheering Bush on and hoping he would do a good job.

Overnight, this became a challenging time for EMILY's List. As we all dealt with the whirling emotions of September 11, fund-raising for EMILY's List all but came to a halt. The political climate instantly became so bad that we had to reassess our goals. Instead of hoping for significant electoral gains, we realized we needed to shore up our women running for reelection.

But the narrative that came out of 9/11 was not the whole story. From the start of the Bush-Cheney administration, there had been tremendous anger on the part of Democrats, because we felt Al Gore really won the 2000 election. In addition, here was this man in the presidency who had campaigned as a "compassionate conservative" yet who was doing the bidding of the right.

What's more, during the summer of 2002, less than a year after the attacks, it became clear that Bush was preparing secretly for war with Iraq—which had nothing to do with the 9/11 attacks. In August, Vice President Dick Cheney asserted that "many of us are convinced that Saddam will acquire nuclear weapons fairly soon."

The first anniversary of the 9/11 attacks presented the perfect opportunity for the Bush administration to push its goal. On Sunday, September 8, 2002, came morning chat-show appearances by Dick Cheney, Colin Powell, Donald Rumsfeld, and Condoleezza Rice on NBC, Fox News, CBS, and CNN, respectively, in which all of them said essentially the same thing. "There will always be some uncertainty about how quickly [Saddam] can acquire nuclear weapons," national security adviser Rice told CNN. "But we don't want the smoking gun to be a mushroom cloud."

The march to war had begun.

AT THE TIME, BUSH'S approval ratings were still sky high, but, as I found out at a meeting in Chicago on October 2, one could already begin to see the first glimmers of opposition. I was at a small dinner with Majority Council members, who had just come from the first big public rally against the war in Iraq. These were respectable-looking, staid women, mostly over seventy years old. One of them told me how much the rally had reminded her of going to antiwar marches in the sixties.

"Oh," I said, "and did it make you want to smoke a joint?"

"Yes," she replied. "And have sex, too."

After I stopped laughing, the women also told me that there had been a marvelous antiwar speech by an incredible Illinois state senator named Barack Obama. The next night, we had a membership meeting with fifty or sixty people, and Barack and Michelle Obama arrived as well. It was the first time I met the Obamas, and I was impressed by how attractive, personable, and smart they both were.

Meanwhile, we had lowered our expectations considerably for the November 2002 elections. This time around, the magical "2" year wasn't so magical. The biggest problem was that congressional districts were redrawn at the state level, where EMILY's List had little or no say. Each state was different, but essentially the governors and legislative leaders were the ones who drew the lines, and —surprise, surprise!—they made deals that protected themselves. Redistricting was the ultimate incumbency-protection act.

Where once I had naively thought we had 435 House seats on the table each election, now I knew that at least 200 seats were solidly Republican and 200 seats solidly Democratic. Certainly those solid Republican seats were off the table for EMILY's List candidates.

That left us marginal seats to try to take from the Republicans, and open seats in Democratic or marginal districts. We were looking at only 20 to 40 House seats where we could realistically add a newcomer to Congress. Initially, we had projected that there would be 7 to 9 seats in play in California alone, but in fact the only marginal seat was that of Gary Condit, a Democrat whose career was effectively destroyed when he was accused of being responsible for the death of an intern with whom he had had an affair. The rest of the fifty-three House districts were either solidly Republican or solidly Democratic.

In addition, GOP-controlled state legislatures were able to realign districts so as to create intramural battles between sitting Democratic legislators. In Michigan, for example, districts had been redrawn so that Democratic representative Lynn Rivers had to run in a primary battle in the same district as Rep. John Dingell, one of the most powerful Democrats in the House as the ranking member of the House Energy and Commerce Committee.

Even Democratic open seats could create serious political challenges. In Illinois, for example, where Rod Blagojevich had given up his seat to run for governor, our candidate Nancy Kaszak was pitted in a primary battle against former Clinton aide Rahm Emanuel, who had access to the resources of the Clinton network. In both the Michigan and Illinois races, it was once again EMILY's List against the party establishment. I decided we had to be true to our mission and help the women candidates, even though both races were extremely difficult to win, and I knew the establishment would try for revenge. In the end, we lost both races.

It was the worst of all worlds: a dreadful political climate, a terrible environment for fund-raising, and an absence of opportunities to add new women. It was at times like these that I especially admired our staff, who buckled down and worked tirelessly. Although we expected the results of the election would be grim, we committed ourselves to protecting vulnerable incumbents in the House and Senate.

We were also particularly worried about Michigan, where Jennifer Granholm was running for governor. Her opponent was Dick Posthumus, Michigan's lieutenant governor and former majority leader of the state senate. Michigan Republicans, realizing the threat EMILY'S List posed, had even tried to change the law to limit the amount of money we could raise for Granholm. Their attempt backfired, however, as EMILY's List members became enraged at this political trick and poured money into Jennifer's race. We launched a massive WOMEN VOTE! program and prayed for success.

ALL IN ALL, 2002 was a terrible election. All our Senate candidates lost, including Sen. Jean Carnahan, running for the full six-year term. Our House members, except for Lynn Rivers, all won reelection, but we added only two new representatives, compared to eighteen newcomers a decade earlier. The only good news came in the form of three great Democratic governors we helped reelect: Kathleen Sebelius in Kansas, Janet Napolitano in Arizona, and Jennifer Granholm in Michigan.

The *Detroit Free Press* provided a moment of joy amid all the disappointing news. In two-inch type, the paper announced Jennifer's victory with a simple statement: SHE's THE BOSS. I couldn't have agreed more.

And, in the aftermath of the election, there was one other positive development. Nancy Pelosi had moved up within the Democratic leadership in the House, and had beaten Steny Hoyer to become House minority leader. She had become the first woman to lead either party in Congress. Her victory was a crystal clear example of women putting women in power. Nancy built her House political base on the Democrats in the California delegation and on the long list of women whom EMILY's List had helped to elect. The combination of those two constituencies was insurmountable.

AFTER THE DISAPPOINTMENTS OF 2002, we at EMILY's List decided to reexamine all of our strategies to win elections. At the state level — the entry level in politics for many women — we were stymied. In fact, when we looked at the data for 2000, we discovered that the percentage of women in state legislatures had dropped for the first time in thirty years. Whether that was a fluke or the beginning of a long-term trend, EMILY's List had to act. Republicans had long been cultivating "farm teams" of young candidates, and it was imperative that we do the same. We had to train hundreds of young women who could start out in local politics and work their way up to the state level, ultimately to play important roles on the national stage.

A year earlier, we had launched our Political Opportunity Program (POP) to recruit, train, and support women to run for state and local offices, and now we began to expand it. Newly hired staff fanned out across the country, building new relationships in the states and encouraging women to run.

For years, we had helped elect women to top offices by raising money, helping them build effective campaigns, and mobilizing women voters to elect them and all Democrats on the ticket. Now, POP became our fourth major tactic. We were committed to developing the next generation of leaders.

I began comparing notes with AFL-CIO political director Steve Rosenthal, one of the smartest and most influential strategists in the Democratic Party. Both EMILY's List and the AFL-CIO were doing significant voter-contact programs with our constituencies, women and union members respectively. Our research showed repeated incidences of small groups of voters getting the same messages from multiple organizations while many voters heard nothing from anyone. In short, our targeting was missing opportunities, and the outside groups were wasting money focusing on the same people. We wanted to fine-tune our work to make sure we were getting the most bang for our bucks.

Another development that brought about a major change in the political landscape for the 2004 elections and beyond was the McCain-Feingold campaign-finance-reform bill, aka the Bipartisan Campaign Reform Act of 2002, which stipulated that political parties were no longer allowed to raise unlimited "soft money." The bottom line was that major donors could no longer pour money into the coffers of either major political party. That left the possibility that millions of dollars would now be funneled through outside political organizations instead.

To take advantage of this changing environment, Gina Glantz, an old friend working for Andy Stern, the president of Service Employees International Union, approached me and suggested that a small group of us get together to come up with a joint strategy. We all had different political missions, but we were united in a fervent wish to defeat President Bush in 2004.

To figure out exactly what we might do, Steve Rosenthal and I met at a Mediterranean restaurant near Dupont Circle, with Andy and Gina, Sierra Club executive director Carl Pope, former Clinton adviser Harold Ickes, and Ickes's business partner Janice Enright. We developed a novel and creative strategy. First, we wanted to create a massive voter-contact program to focus on critical battleground states. This program we named America Coming Together (ACT). Steve Rosenthal would be CEO and would put together the political operation, while I would serve as president and would raise the money. Second, we wanted to create a coalition of progressive nonprofit organizations to run voter-contact programs. This became America Votes. Our goal was to coordinate our efforts in order to significantly boost the efficiency and cost-effectiveness of voter contact. Cecile Richards, Ann's daughter, would leave her job working for Nancy Pelosi and run America Votes. We would use data collected by ACT to help target voters and send them messages from the organizations they would most likely respond to. And finally, Harold Ickes wanted to create a fund to broadcast messages in support of the Democratic presidential nominee on radio and television. This became the Media Fund.

This three-pronged strategy was sound politically but lacked a critical ingredient: money. The group was both surprised and excited when we learned that billionaire George Soros was interested in investing significant dollars in a plan to defeat the Bush-Cheney ticket. Eventually, he decided to help fund ACT and convinced some of his friends, including fellow billionaire Peter Lewis, the chairman of the Progressive Insurance Company, to join him. When we announced we had raised $20 million to launch ACT, the political world began to take notice. Harold and I hit the road to raise the rest of our funds.

ACT was able to fund an army of more than forty-five thousand paid canvassers and twenty-five thousand volunteers, who made more than twelve million phone calls to targeted voters and deliveries of political literature to eleven million targeted doorsteps in battleground states. It was an enormous job for me, and I had to take a partial leave from EMILY's List to do it. But within this Democratic coalition, EMILY not only had a seat at the table, we were a recognized leader.

IN MARCH 2003, the Iraq War began. To tens of millions of Americans, President Bush instantly became a heroic wartime president who, presumably, would be invulnerable when he came up for reelection the following year. Just two months later, Bush, clad in fighter-pilot regalia, delivered his famous Mission Accomplished speech on the deck of the U.S.S. *Abraham Lincoln* aircraft carrier, asserting that the war was over.

Bush's victory celebration was premature, of course — by more than eight years, it turned out. And when the war continued and it became increasingly clear that the WMDs that had been the pretext for the invasion did not exist, one ugly element after another in the Bush-Cheney administration emerged. We were becoming a national-security state. We were torturing prisoners in Abu Ghraib and Guantánamo Bay. Antiwomen policies, clad in the cloak of Christian evangelicalism, were becoming pervasive throughout the administration. As 2004 drew near, the urgency of unseating Bush grew and grew.

As I crisscrossed the country raising money for ACT and EMILY's List, I would periodically check in with the POP staff to see how they were progressing with their efforts in recruiting, training, and supporting women as they sought political office. By

September 2003, we had more than five hundred potential or declared candidates in seven states, and fifteen training seminars in the works.

POP was helping us develop new political relationships in the states. In New York, for example, state senator Liz Krueger, herself a beneficiary of help from POP in 2002, suggested to Kate Coyne-McCoy, our East Coast POP director, that we work with David Paterson, then minority leader of the state senate, who was building a multielection strategy to take back the Senate from Republican control. In New York, the governor, speaker of the assembly, and the senate majority leader typically draw boundary lines in redistricting. Democrats already held the state assembly and were expected to elect Eliot Spitzer as governor in 2006. If Democrats could take over the Senate, we would win the redistricting trifecta after the 2010 census. And with friends in these positions, EMILY's List could create some new opportunities for women in both state and federal offices. We were in!

Well aware that we helped only women candidates, Paterson told Kate that he was concerned that male Democratic senators might lose their seats because of how their districts had been redrawn, and asked whether POP would provide training sessions for them. We branched out from our usual plan and agreed to train male state senators who hadn't run in a tough race for years.

BY THE SUMMER OF 2004, EMILY's List and our allies were working at a fevered pitch to defeat Bush and elect Democrats. ACT, under Steve Rosenthal's leadership, had thousands of paid canvassers going door-to-door, talking with voters about their concerns and entering the data on their Palm Pilots to be added to a massive and growing database. America Votes was organizing, through its member organizations, volunteers to go door-to-door on weekends. EMILY's List had recruited a strong roster of candidates and was helping them build effective campaigns, win their primaries, and head toward the final lap. Joe Solmonese, executive director of EMILY's List, was keeping us on track as I flew back and forth across the country raising millions of dollars.

No race better illustrated how all these new components coalesced than Gwen Moore's 2004 congressional race in Milwaukee, Wisconsin. An extraordinary woman by any measure, Gwen was a dynamo in the Wisconsin state senate who had won great loyalty from her inner-city Milwaukee constituency. She was a genius at grassroots campaigning but needed significant help expanding her base to a larger and broader congressional district. A single African American mother who had been a recipient of welfare and food stamps when she graduated from Marquette University, in 1978, Gwen traversed chasms of race, class, and gender that were as wide as those faced by any politician I've known.

What she achieved was phenomenal. Starting out as a community organizer in Milwaukee, Gwen soon realized that one of the problems in her community was that banks wouldn't lend money to anyone in her largely black neighborhood. Rather than complain, Gwen helped establish the Cream City Community Development Credit Union in Milwaukee to offer grants and loans to low-income residents who wanted to start small businesses. Gwen was a problem solver.

In 1988, Gwen won a seat in the Wisconsin State Assembly. Then, three years later, she gained national recognition when she called for an investigation into the Milwaukee Police Department after it repeatedly ignored calls about a fourteen-year-old Laotian boy who was seen naked and bleeding running in the streets in Gwen's neighborhood. Asserting that the police dismissed the reports because they came from African American women, Gwen noted that officers returned the boy to the apartment of a white man without even bothering to check the man's background. The white man turned out to be none other than Jeffrey Dahmer, the serial killer who raped, murdered, and dismembered seventeen boys. Remains of the young Laotian were later found in Dahmer's apartment.

In 1992, the next year, Gwen became the first African American

woman ever elected to the Wisconsin State Senate. For the next
twelve years, she focused on issues that were all about helping the
urban poor. She was beloved in her district. Then, in 2004, she decided to run for Congress from Milwaukee to fill a seat that opened
up when Rep. Jerry Kleczka announced his retirement.

Kleczka's seat was solidly Democratic, so the real challenge was
the primary. There, Gwen faced several problems. First, redistricting in 2000 meant that Gwen's black constituencies now shared the
same congressional district with multiethnic, working-class South
Side Milwaukee voters. That meant Gwen needed to broaden her
support with white voters.

Second, Gwen faced two significant opponents in the primary:
Matt Flynn, a hotshot defense attorney and former chair of the Wisconsin Democratic Party; and Tim Carpenter, a gay state senator
who was a long shot. The Democratic establishment considered
Flynn to be the strong favorite.

Even though Gwen announced her candidacy just after the seat
opened up, she immediately ran smack-dab into the curse of the old
boys' network: Flynn had already lined up an endorsement from the
retiring Kleczka and the tacit support of Gov. Jim Doyle and longtime Democratic representative David Obey, and soon amassed
more campaign funds than all his opponents combined.

"When I started calling people to see if they would support me,
they were already committed to Flynn," said Gwen. "How could
they have been committed when the announcement [of Kleczka's
retirement] was just made yesterday? The answer is that Flynn already knew Kleczka was going to retire."

Hoping to head off a battle with her, Flynn urged Gwen to stay
in the state senate, where they would be able to work together. Be
sensible, he said.

But Flynn didn't know Gwen. "I just got right into his face,"
Gwen said. "You could smell my breath. And I said, 'Look deeply
into my eyes and tell me what you see. Do you see the aspirations of

a generation of people, or do you see a sensible person?' He thought I was crazy."

Even though it meant giving up her state senate seat, and with no campaign funds or support from the party establishment, Gwen was off and running.

Gwen had been trained by our POP program, but she still had only a vague familiarity with EMILY's List. For the most part, she dismissed us as a bunch of rich white ladies. "I had heard of EMILY's List," she said. "But I just knew they weren't going to help me. I had heard that they didn't help black people."

But that was precisely where Gwen was wrong. In fact, more than one-third of the women we helped elect were women of color — and we were exactly what Gwen needed if she was to expand her base. So, we immediately assigned Jennifer Pihlaja, a young staffer at EMILY's List, to work with Gwen.

At first, Gwen was standoffish. When Jen told her we would support her, she responded, "I know, you're the people that are gonna make me swim across the English Channel, and when I get to the other side, you'll be there with a warm towel and hot chocolate."

Jen told Gwen she was exactly what we were looking for, a committed pro-choice Democratic woman who would be an excellent member of Congress. And wait until she saw what EMILY's List would do to help!

Gwen was a fascinating candidate. In the state senate, she had become a powerful voice who had real fire in her belly. She gave startlingly candid speeches on the floor of the statehouse using her own life as an unwed mother and someone who had been assaulted as a child as an example. "This woman is fantastic," said Jen. "She's one of the smartest people I know. She has a photographic memory. She's crafty. Starting a credit union was typical of the way Gwen thinks and acts: we should have this, so she did it. She was terrific on issues that were important to her district, such as food stamps and housing."

The problem was that none of those issues brought in political money. I've never seen a candidate who had such a small potential fund-raising base — and Gwen was almost proud of it. "One of her supporters gave her a box of pennies, and I saw Gwen literally counting the pennies," said Jen. "She wanted to make the point that that contribution meant as much to her as all the hours spent raising money on the phone from white liberals." So, suddenly you had an understated, mild-mannered young white woman from the Midwest — Jen — trying to tell this unstoppably powerful black woman what she needed to do. It wasn't easy.

When it came to debating, Gwen's rhetoric was not exactly geared to the affluent white, liberal donors she needed. As Gwen put it, her style was "street." "I know how to argue," she said. "I know the facts. But before EMILY, my style was something like this." To demonstrate, Gwen slowly took off her glasses and looked directly at me as if I were her opponent in a televised political debate. "Kiss my ass, motherfucker," she said. "I'm gonna tell you how it is . . ." Then, she burst into laughter.

Of course, Gwen was putting us on. She was much too smart to blow up her own campaign. But, just in case, we made sure Gwen's rhetorical skills were refined somewhat by the time the election was over.

Amazingly, it all began to work.

EMILY's List legally couldn't do the actual work of the campaigns — fund-raising, organizing, and the like — but we had a continuing interest in assessing our races. Of course, a critical part of that process was advising staff, many of whom were new to their jobs, on what was working and what wasn't.

So, we sent Becca Sharp and Dave McGonigle to monitor Gwen's field operations and her get-out-the-vote process. In the end, we sent so many staffers from EMILY's List that Gwen used the name Emily as a prefix for all of them: Emily Jen, Emily Dave, Emily Heather, and so forth. I was proud to be known as Emily Ellen.

As the September primary approached, I suggested to Steve

Rosenthal that it might be a good idea to use Gwen's primary as a dress rehearsal for ACT's get-out-the-vote operation in Milwaukee. Steve, of course, was way ahead of me and was already planning to test out the new ACT political machine for the September 15 primary.

Gwen was a powerful and much-loved candidate. But it was also true that our relationship with her was a terrifically productive political marriage. In the last reporting period before the September primary, Gwen raised $354,000 — four times as much as Matt Flynn, and more than $200,000 of that came from our members. In late August, WOMEN VOTE! went on TV with an issue ad praising Gwen for her work on education. We were at her side every step of the way.

When the votes came in on September 15, Gwen earned her win in what can only be described as a colossal landslide, winning 64 percent of the vote to just 25 percent for Flynn. It was hard to remember that just a few months earlier, Flynn had been a prohibitive favorite.

WHEN NOVEMBER 2004 came around, once again we had a pretty impressive election cycle. Our POP program had trained 1,600 current and prospective candidates in forty seminars in twenty-nine states. Claire McCaskill almost won her race for governor of Missouri, but, alas, she was dragged down by John Kerry's poor showing at the top of the ticket. However, we helped elect many potential new local and state leaders, including Kamala Harris, who became the first African American district attorney in San Francisco; Houston city controller Annise Parker; and Robin Carnahan, the daughter of Mel and Jean Carnahan, who ran for secretary of state in Missouri. Overall, we helped elect 140 women to state and local offices from twenty-eight states.

Every single EMILY's List candidate seeking reelection won, including Senators Barbara Boxer and Patty Murray, both of whom had been targeted by Republicans. Christine Gregoire was elected governor of Washington, and Gov. Ruth Ann Minner was reelected in Delaware.

We added five new women to the House of Representatives, including Gwen Moore, who became the first African American elected to Congress from Wisconsin, and an up-and-coming young Floridian named Debbie Wasserman Schultz.

There was one other aspect of the election that I particularly loved. When Gwen decided to run for Congress, we realized, of course, that she was vacating her seat as state senator, and that was an opening we should not overlook. So, our POP program went into high gear, and we suggested to state representative Lena Taylor that she run for Gwen's state senate seat. She did.

But now Lena was vacating *her* seat, so we needed yet another candidate. We told Lena that one of her first jobs was to recruit a woman to run for her state representative seat. Lena recruited a social worker and university professor named Tamara Grigsby to run for the state legislature. When Gwen, Lena, and Tamara all won, we referred to it as our Wisconsin trifecta. Three African American women were now representing Milwaukee in the Wisconsin state legislature, the state senate, and the House of Representatives.

THE TOP OF THE TICKET was a different story, but even there, EMILY's List more than pulled its weight. With the memory of the Bush-Gore catastrophe in Florida four years earlier still fresh in our minds, we were determined that Palm Beach County, where the highly confusing so-called butterfly ballots produced an unexpectedly large number of votes for conservative third-party candidate Patrick Buchanan, would not doom Democratic nominee John Kerry as it had doomed Al Gore.

EMILY'S List told our American Votes partners that we wanted to take on the challenge of Palm Beach County in 2004. And so, Air EMILY, a massive get-out-the-vote operation, went into action. More than 1,300 EMILY supporters flooded Palm Beach. An entire parking lot was packed with vans we had rented, each of which had voter lists, boxes of material, and water and fruit for volunteers. Hundreds of people were divided into teams to get out the vote. Our operation was hugely successful, with voter turnout in Palm Beach County reaching an astounding 75 percent.

We had made great strides during the 2004 cycle in building a political operation for years to come. Our POP program had really gotten into the nitty-gritty at the state and local levels, training women to run and preparing them for the future. Our database skills were becoming increasingly sophisticated in determining which voters needed more contacts to convince them to go to the polls for the Democrats. Because we had better data, we were better able to target and mobilize these women through WOMEN VOTE! We had trained hundreds of young people, we had developed an unprecedented ground campaign in Florida, and we now had one hundred thousand members of EMILY's List — all of which boded well for the future. As political analyst Charlie Cook put it, "Democrats, chiefly through America Coming Together, mounted what was not only the most sophisticated get-out-the-vote operation in the party's history, but it was probably the best field work by a factor of at least ten."

But the bottom line was that when all the votes were counted, the Republicans had retaken the Senate, and, at the top of the ticket, the Democrats once again suffered a bitter, heartbreaking loss, with Bush eking out a win over John Kerry by 35 electoral votes. That meant the GOP controlled the White House and both houses of Congress.

Kerry's loss to Bush sent a wave of anger and discontent through the donor community that supported ACT and our strategy to win in 2004. Donors withdrew their support, erroneously concluding that ACT was a failure. Steve, Harold, and I were more than willing to return our normal lives and decided to disband ACT.

But our work in 2004 created continuing, important changes in the Democratic political coalition. The building of an unprecedented database of voter information — all those Palm Pilots — took

Democratic campaigns into new, technology-based ways of operating. Gone were the days when campaigns relied solely on paid media to convince voters. Supporters and donors alike now realized that personal, targeted voter communication was a critical element of success.

EMILY's List emerged from 2004 in a new position in the Democratic coalition. Though we had previously operated largely on our own, through America Votes we learned the impact of building coalitions to join together in targeted, efficient voter contact. The coalition realized that our expertise with women voters and the massive resources we could mobilize made us a force to be reckoned with, not just for our women candidates but for Democrats at all levels. We were no longer, in the words of the character Toby Ziegler on NBC's *The West Wing*, the "girls' group with the funny name." We were major players in the Democratic coalition.

Exhausted yet proud, I joined the staff in a period of rest and reflection. But, as always, that didn't last long. Before you knew it, we were focused on a new mission for 2006. It was time to take back the House of Representatives — and we needed to switch only sixteen House seats to do it.

Fifteen

MADAM SPEAKER

THE ATTACKS OF 9/11 and the wars in Afghanistan and Iraq had so dominated George W. Bush's first term that people often forget that his presidency had been off to a weak start before these events took place. Indeed, in the first eight months of Bush's administration, his disapproval numbers nearly doubled, to nearly 40 percent, and his approval ratings were on the verge of sinking below 50 percent. The trend line was clear: the American people thought he was doing a lousy job. It was only after 9/11 that the entire country rallied behind him.

In his second term, however, the patriotic fervor behind the Iraq War had dissipated. Millions of Americans now realized that the pretext for war — Iraq's WMDs — was a lie. They had seen Bush declare, "mission accomplished," and realized that this was a lie as well. He had created a quagmire that was taking thousands of American lives and costing trillions of dollars, and he didn't have a clue as to how to get out of it.

At the same time, Bush seemed oblivious to how much his support had eroded. Even though he barely squeaked by in the 2004 elections, he boldly asserted that he had "earned capital in the campaign, political capital, and now I intend to spend it." Specifically, the top item on his agenda was an attempt to privatize Social Security. Bush was taking aim at the most sacred tenet of the New Deal, and he didn't seem to realize he was running out of ammunition. In January 2005, just after the new Congress was seated, Bush embarked on a long series of nationwide tours to mobilize public opinion for his proposal. But the more he argued for his plan, the more unpopular it became. By early summer, it was dying.

Then, in late August 2005, Bush's woes were compounded exponentially when Hurricane Katrina demolished New Orleans. Hundreds of thousands of lives were devastated, and all over the country people leaped forward to help. But instead of bringing us together, the voice of "compassionate conservatism" came forth with the most cavalier response imaginable, famously telling FEMA director Michael Brown, "Brownie, you're doing a heckuva job." FEMA's response to the disaster was itself a disaster.

The upshot of these catastrophes was that the country wanted change, and Democrats were suddenly in a position to gain real ground. Once unassailable Republican incumbents suddenly seemed vulnerable. So, as the 2006 midterms approached, EMILY's List recruited a top-tier roster of women candidates to go after every possible "take-back" seat imaginable. In upstate New York, we found a young attorney named Kirsten Gillibrand to take on Republican incumbent John Sweeney in the House. It was a heavily Republican district, but Sweeney, once a moderate, had leaned so far to the right that he was out of step with his district. Patricia Madrid in New Mexico, Gabrielle Giffords in Arizona, Lois Murphy in Pennsylvania, and Diane Farrell in Connecticut were strong challengers in their respective districts. In Illinois, Tammy Duckworth, a veteran of the Iraq War who had lost both her legs and shattered an arm in combat, was running for the seat of retiring representative Henry Hyde, a Republican.

In the Senate, two key races showed how essential our POP program could be. One was in Minnesota, where Democrat Mark Dayton had announced he would not run for reelection. Our political director, Martha McKenna, was on it in a flash and called Ann Liston, the regional director of POP, to find a replacement candidate.

Ann immediately suggested Amy Klobuchar, the forty-four-year-old county attorney in Minneapolis's Hennepin County.

A 1982 Yale graduate who received her law degree at the University of Chicago Law School, Amy had been a partner at Dorsey & Whitney, a large Minneapolis firm, and served as legal adviser to former vice president Walter Mondale, who was also a partner at the firm. She had two potential primary challengers: Patty Wetterling, a former House candidate whom we had supported two years earlier in a losing congressional race; and Mike Ciresi, a wealthy trial lawyer.

Amy entered the race at a feverish pitch, raising money and putting together a strong campaign with our counsel. It quickly became clear that she was the strongest of the two women in the race and that her best chance of winning was to avoid an expensive, negative primary, so we endorsed her early, on September 29, 2005. Not long afterward, Wetterling told Martha McKenna that it was as if someone stepped on a garden hose; all the money stopped flowing into her campaign. In January 2006, Wetterling withdrew from the race and threw her support to Amy. Similarly, after assessing his prospects, Ciresi decided against running. A textbook early-money strategy had enabled us to avoid a divisive and costly primary battle and put Amy in a strong position for the general election.

Our other key senatorial race that season was in Missouri, where a vulnerable Republican incumbent, Jim Talent, presented a take-back opportunity in a seat that had changed hands twice in the previous six years, both times by narrow margins. This was the seat that the late Mel Carnahan had won posthumously in 2000 but that Republicans regained two years later, when Talent beat Jean Carnahan, who had been appointed to the Senate seat.

For the race against Talent, we were excited about the prospects of state auditor Claire McCaskill, whom we had supported in her narrow loss in the gubernatorial race of 2004. The more time I spent with Claire, the more I respected her ability to dispassionately assess her flaws, both political and as a candidate, and to address them. Back in the 2004 gubernatorial race, she had lost badly in the rural areas outside of Saint Louis and Kansas City, so this time she sent her mother there as an ambassador. Claire took charge of decision making on ads so she would not be blindsided. She opened up and became a warmer, more authentic candidate.

In addition, we knew we would face a tough fight in Michigan, a key battleground state and the site of Debbie Stabenow's narrow victory in 2000, because the economy was disintegrating as the auto industry collapsed. Both Debbie and Gov. Jennifer Granholm were up for reelection in 2006, and we didn't want them to become the victims of voters' discontent. Corporate lobbyists, especially from the pharmaceutical industry, were eager to oust Debbie for her consumer-friendly work on prescription-drug costs, and to that end the Republicans put up both Mike Bouchard, a veteran police officer inspired to run for office by the September 11 attacks, and evangelical pastor Keith Butler as their candidates. Also in Michigan, Republican gubernatorial candidate Dick DeVos, the billionaire scion to the Amway fortune, said he would spend as much as $100 million of his own money to take on incumbent Jennifer Granholm.

HILLARY CLINTON WAS ALSO up for reelection to the Senate in New York in 2006, but because she was such a shoo-in, pundits busied themselves obsessing about a larger question: Was she was going to run for president in 2008? As early as 2003, Clinton pollster Mark Penn had been secretly assessing Hillary's presidential appeal. Hillary decided not to run in 2004, but by 2005 the presidential chatter had begun. In October, lawyer and author Susan Estrich published *The Case for Hillary Clinton,* advocating that Hillary run for the presidency. In February 2006, ABC News was asking whether Hillary, "simultaneously one of the most revered and reviled people in the country," was running for president.

By this point, the answer was that Hillary was seriously considering a 2008 run but was keeping it quiet. Throughout the rest of the year, the subject provided fodder for political analysts and TV chat shows day after day after day.

The good news was that finally, twenty years after we had founded EMILY's List, a number of women had become viable candidates for the most powerful jobs in the land. It's worth remembering that when we had started, in 1985, the Democratic Party had never in its entire history elected a woman senator in her own right. The political establishment had made it so difficult for women to get elected that no women had the experience and credentials necessary to run for higher offices. But now, there were enough women senators, representatives, and governors that voting for a woman for high office was no longer a novelty. Voters knew what women politicians looked like and how they acted. Now, they were less likely to judge women based on gender stereotypes and more likely to assess their character and their positions on issues.

Moreover, a considerable number of women had so much experience, such substantial track records, that they were serious contenders at the highest levels of government. Nancy Pelosi had risen to the top of the party leadership in Congress, and was in line to become Speaker of the House if the Democrats won back the House. And Hillary had such an impressive résumé regarding veteran's issues, human rights, foreign affairs, children, families, health care, and women's rights that she was now widely perceived as presidential material. "From her earliest days as a law student," said Judy Lichtman, who had known Hillary since the seventies, "she has been a leader in addressing the concerns of women and our families and people who cannot protect themselves and represent themselves."

As First Lady, Hillary had helped create CHIP, the Children's Health Insurance Program, which at the time was the largest single investment in health care since Medicare. She put the Taliban's brutal treatment of women high on the administration's agenda. And during her trip to Beijing for the World Conference on Women in

1995, Hillary stood up against both internal administration pressure and pressure from the Chinese to declare that "it is no longer acceptable to discuss women's rights as separate from human rights . . . If there is one message that echoes forth from this conference, let it be that human rights are women's rights and women's rights are human rights, once and for all."

As senator, Hillary worked to enact legislation to protect kids from drugs and to dramatically expand CHIP. Hillary led the charge in the Senate to strengthen equal-pay laws by introducing the Paycheck Fairness Act to strengthen penalties against wage discrimination. And she was an original cosponsor of the Lilly Ledbetter Fair Pay Act, the first bill signed by President Obama in 2009.

Hillary fought to expand family planning to low-income women. She backed legislation requiring health-insurance companies to cover contraception. With Patty Murray, senator from Washington State, she waged a three-year battle to get the Food and Drug Administration to make the morning-after pill available. She was a strong supporter of tax cuts for the middle class. And, long before Elizabeth Warren was elected senator from Massachusetts, Hillary advocated ending loopholes that allowed investment managers at hedge funds and private-equity firms to dramatically lower their income tax.

But, after more than a decade on the national stage, Hillary also carried with her considerable baggage. As the *Washington Monthly* put it in a 2005 article about Hillary's White House prospects, "You know the rap: She's too liberal, too polarizing, a feminist too threatening to male voters. Too much baggage. Too . . . *Clinton.*"

And so, the seeds were being sown for an epic political drama about whether and when we might have the first woman president of the United States.

NANCY PELOSI'S ASCENT as Speaker was more clearly defined and more immediate. After knowing her for nearly twenty years, I had come to see Nancy as a brilliant strategist who was savvy, tough, and pragmatic — terrific qualities for a Speaker of the House. A natural pol if ever there was one, Nancy was on a first-name basis with two hundred members of Congress by the time she took office, in 1987. Once she was seated in the House, she hit the ground running, pushing for federal money for AIDS research and for human rights in China. She joined the powerful Appropriations and Intelligence Committees, became one of the first members of the Congressional Progressive Caucus, and began her ascent within Congress.

Having come to the House when it was almost entirely male dominated, Nancy had the strategic acumen necessary to forge alliances with key people who could help her get things done. "The legislature is not for the faint of heart," said Rosa DeLauro (D-CT). "No one gives you anything in this institution. You've got to take it. And Nancy had an unbelievable number of qualities that enabled her to do that. She's masterful at cultivating strategic relationships."

In addition to putting together the daunting "DeLowsi" triumvirate with Rosa DeLauro and Nita Lowey on the House Appropriations Committee, Nancy shrewdly forged a highly productive alliance with Jack Murtha (D-PA), knowing that, as the chair of the powerful House Defense Appropriations Subcommittee, he was in a position to deliver. "In those early days, getting a hundred million dollars for breast cancer research was a big deal," said Nancy. "We looked at all the competition for funding, and we realized you had to be careful. If you wanted to get more money for AIDS or breast cancer research, the Republicans would take it out of children's education or the Occupational Safety and Health Administration."

One exception, however, was defense appropriations. The military budget was sacrosanct to Republicans. "So, we went to Jack Murtha when we wanted funds for AIDS or breast cancer," Nancy said. And Jack Murtha came through.

That created the unusual situation wherein Joanne Howes, one of EMILY's Founding Mothers who was also a lobbyist for the National Breast Cancer Coalition, ended up getting a call from a high-ranking general saying he had millions of dollars to spend on breast cancer research and asking her for advice on what to do with it. She steered him in the right direction.

It was a circuitous route to line up funding for medical research for women, but it worked. Over the years, Murtha secured more than $2.3 billion in funding for breast cancer research.

Nancy used similar stratagems to get funding for Housing Opportunities for Persons with AIDS, to accelerate development of an HIV vaccine, and to expand Medicaid for people with HIV. And she was a powerful force in passing bills to reform health care, raise the minimum wage, and protect women's reproductive health, among other issues.

Because her California seat was one of the safest Democratic seats in the country, Nancy did not need EMILY's List's support in her own campaigns. But when it came to her moving up the ladder within the House, EMILY's List was crucial. EMILY's List had almost quadrupled the number of Democratic women in the House, and that gave Nancy a formidable political base. So, in 2001, when she was running for minority whip, she was able to get support from most of the thirty-two Democratic California representatives and forty-two Democratic women in the House. With overwhelming support from those two large blocs, Nancy beat out Steny Hoyer to win the party's number-two spot in the House. Then, in 2002, when Dick Gephardt resigned as minority leader to run for president, Nancy moved up to take his place, beating Harold Ford Jr. of Tennessee.

As minority whip and as minority leader, Nancy made party discipline central to her way of doing business, and that meant closely monitoring the "favor bank" that is part of life on Capitol Hill. "She understands you put in, you put in, you put in, and then, at some point, you say, 'I need a favor,'" said Rep. Jim McDermott (D-WA), a friend. "She won't say, 'Vote this way.' She's very realistic. She'll say, 'You are free to do what you want.' But you can be sure she'll remember if you don't do the right thing." And when that happened, Nancy was tough. According to Rep. Anna Eshoo (D-CA), also a friend, when a fellow Democrat told Nancy, in effect, "I'm sorry, I can't be with you," Nancy almost always had an icy reply: "We can't be with you either."

But her toughness had paid off. And now, if the Democrats took over the House, becoming Speaker would be a mere formality.

AS THE ELECTIONS APPROACHED in 2006, Republicans were dogged by one scandal after another. Disgraced superlobbyist Jack Abramoff pleaded guilty to charges of tax fraud, tax evasion, and conspiracy. Rep. Duke Cunningham (R-CA) was convicted of taking more than $2 million in bribes. Texas representative Tom De-Lay, the daunting House majority whip known as "the Hammer," resigned rather than run for reelection, because of investigations into the financing of his political-action committees. Florida representative Mark Foley resigned after it was revealed he had sent suggestive e-mails and sexually explicit instant messages to teenage boys who were congressional pages. A U.N. report on detainees at Guantánamo concluded that the rights of the prisoners there were violated and that, in some cases, U.S. treatment constituted torture.

Most important of all, Bush had more than forfeited his political capital. In the aftermath of Hurricane Katrina, New Orleans remained a shameful, festering sore, a great American city in ruins, with tens of thousands of people living in trailers, their fates uncertain. And now that the public had turned against the war, everything that had once made Bush and the Republicans so strong — their discipline, their insistent message that only they knew how to fight the war on terror — now made them seem weak, in denial, out of touch. Iraq had devolved into a civil war, and there were no signs that the United States could extricate itself.

By October 2006, Bush's approval rating had hit rock bottom. Polls by *Newsweek,* Pew Research, and CBS News–New York Times all had his approval rating in the low 20s, a drop of nearly 70 percentage points from his peak. The Republicans were petrified.

Meanwhile, now that Democrats finally realized they could win, money poured in to EMILY's List so fast that our daily deposits frequently exceeded $500,000. Our WOMEN VOTE! director, Maren Hesla, began a massive get-out-the-vote operation that was carefully coordinated with political allies.

When you studied the numbers, it looked like the Democrats could win back at least one of the chambers. To take over the House of Representatives, we needed to pick up fifteen seats, and forty were in play. That was feasible. Winning back the Senate required us to pick up six seats, which was more difficult. But the prospects were bright enough that Democrats became giddy about what committee chair they might win and what agendas they might pursue. "I've acquired a lot of new friends this year," said Rep. Barney Frank (D-MA), who was in line to become chair of the House Financial Services Committee. "And I haven't gotten any nicer."

For her part, Nancy Pelosi prepared by handing out cards printed with all the things she wanted to accomplish in her first one hundred hours if the Democrats won. Called "6 for '06," her agenda called for hiking the minimum wage, securing critical funding for stem-cell research, providing affordable health care, achieving energy independence, making college affordable, and leading more forcefully on national security.

"We know there's a hurricane coming and it's going to hit the Republicans in November," said political analyst Charlie Cook. "We're just trying to figure out how big this thing is."

The answer? It was huge.

In fact, on November 7, 2006, Democrats dealt Bush the biggest midterm setback since 1994, taking the House of Representatives 233 to 202 seats, and the Senate 51 to 49.

In the Senate, four pro-choice women were elected, including two newcomers, Amy Klobuchar and Claire McCaskill. Sen. Maria Cantwell, who had been targeted by Republicans, won reelection in Washington, as did Hillary Clinton, easily beating Republican John Spencer by a margin of more than two to one. Governors Janet Napolitano and Kathleen Sebelius, both of whom had faced tough gubernatorial battles in Republican states, won reelection. Maren Hesla's huge WOMEN VOTE! effort in Michigan helped save both Sen. Debbie Stabenow and Gov. Jennifer Granholm, who both won reelection. We added seven new Democratic women to the House, bringing the total to fifty. And in local and statewide races, our POP program helped elect 171 women and win Democratic control of six state legislatures.

It was the first time in American history that Republicans failed to win a single House, Senate, or gubernatorial seat previously held by a Democrat. And Nancy Pelosi, as sitting minority leader, was about to make history by becoming the first woman Speaker of the House.

LATE IN THE MORNING on January 4, 2007, I made my way through security and followed the growing crowd to the visitors' gallery of the House of Representatives in the United States Capitol. Nancy Pelosi was being sworn in as the first woman Speaker of the House — two heartbeats away from the presidency.

I had a front-row seat. What made this historic day particularly thrilling to me was knowing that EMILY's List had played a key role in making it possible. When we began EMILY's List, we had dreamed of giving women political power, and now Nancy was about to become the most powerful female elected official in American history.

I took my seat on the balcony overlooking the Democratic end of the House chamber. The gallery was filled to the rafters with friends, relatives, and staffers of the 435 representatives being sworn in. Behind me sat Mark Kelly, the forty-two-year-old astronaut who was there to see his fiancée, Gabrielle Giffords, begin her first term as representative for Arizona's Eighth Congressional District. We had supported Gabby as an EMILY's List candidate, so I introduced myself to Mark and we chatted amiably. Everyone was talking to one another, waving to friends and family. These were difficult, polarizing times in America, but we had taken back the House, and it was a good time to be a Democrat. The atmosphere was electric, celebratory.

Looking down at the House gallery, I could literally see what we had accomplished at EMILY's List after more than twenty years. For decades, the only colors on the floor of the House had been those of the dark suits worn by the men in Congress. I once told Dick Gephardt, the former House Democratic leader, that EMILY's List would bring some color to the House. Now, a glance at the House floor showed that we had done just that. There were the vibrant red, yellow, pink, and blue women's suits and dresses sprinkled throughout the crowd. There were now seventy-three elected women on the floor of Congress. Nearly three-quarters of them—fifty-three—were on the Democratic side of the aisle, and each one of those women exemplified an EMILY's List story.

Retaking the House was a great victory, of course, but it was even sweeter given that this was our twentieth year. During that span, we had seen women go from being bit players to becoming major figures. At last, we could truly say that women had real political power. Louise Slaughter (D-NY) took over as the first woman chair of the House Rules Committee, Nydia Velázquez (D-NY), the first Latina to chair a full congressional committee, took over Small Business, and Juanita Millender-McDonald (D-CA) ran the House Administration Committee. Nita Lowey and Rosa DeLauro chaired subcommittees of the House Appropriations Committee. DeLauro, who had served as EMILY's List's first executive director before moving on and winning her seat in Connecticut's Third District, was an especially sweet sight. And in the Senate, Barbara Boxer chaired the Senate Environment and Public Works Committee, and Dianne Feinstein the Senate Rules and Administration Committee.

I saw two women from Wisconsin in the House who both represented historic firsts: Gwen Moore, the first African American elected to Congress from Wisconsin; and Tammy Baldwin, the first openly gay woman elected to the House. From California, there was Anna Eshoo, our first California win; Doris Matsui, our most recent; and Loretta Sanchez and Linda Sánchez, the first "sister act" elected to Congress. I could spot so many wonderful, successful, courageous women. So many amazing stories. So many memories. So much success.

THE CEREMONIES BEGAN at noon, at which point both houses of Congress were sworn in. On the other side of the Capitol, in the Senate, Vice President Dick Cheney was swearing in the senators who had won that November. He probably would have preferred to be doing something else, but, as president of the Senate, he was charged with administering the oath to no fewer than six EMILY's List candidates who had been elected or reelected in 2006 — Dianne Feinstein of California, Hillary Rodham Clinton of New York, Maria Cantwell of Washington, Debbie Stabenow of Michigan, and two newly elected senators, Claire McCaskill of Missouri and Amy Klobuchar of Minnesota. Counting incumbents who were not up for reelection, we now had a total of eleven Democratic women in the Senate.

From my balcony seat, the House swearing-in ceremony was quite simple. Just after noon, 435 congressmen and -women, plus five nonvoting delegates from Guam, Puerto Rico, American Samoa, the U.S. Virgin Islands, and the District of Columbia, rose to their feet and took the oath of office en masse. The representatives were now called upon to vote for their new leader, the Speaker of the House. And since the Democrats had taken over the House in the midterm elections, the winner was a foregone conclusion.

Now, Nancy Pelosi was in a position to enjoy her moment in the sun. As the motion was made to vote for Speaker of the House and as the representatives called out their votes one by one, Pelosi surrounded herself with her six grandchildren, cradling her youngest grandson, who had been born after the November elections. One of the other young ones kept swinging a microphone. All of them seemed oblivious to the significance of what was going on.

In the end, in a straight party-line vote, Pelosi beat Republican John Boehner, the outgoing Speaker, 233–202. "Whether you're a Republican, a Democrat, or an independent, today is a cause for celebration," Boehner said graciously before handing the gavel over to Pelosi.

Still accompanied by her grandchildren, Nancy invited the children and grandchildren of other representatives to surround her as well. "I accept this gavel in the spirit of partnership, not partisanship," she said, waving the Speaker's gavel. "This is an historic moment — for the Congress, and for the women of this country. It is a moment for which we have waited more than two hundred years. Never losing faith, we waited through the many years of struggle to achieve our rights. But women weren't just waiting; women were working. Never losing faith, we worked to redeem the promise of America, that all men and women are created equal. For our daughters and granddaughters, today, we have broken the marble ceiling."

There were tears in my eyes. We had come such a long way.

It was a day to celebrate — while knowing that tomorrow would come. We still had a long way to go.

Sixteen

DOWN TO THE WIRE

ON THE NIGHT OF January 4, 2007, hours after Nancy Pelosi was sworn in as Speaker, I attended a dinner at the Italian Embassy in her honor, at which I ran into Barack Obama. I had known him since 2002, when he was an Illinois state senator and had given his famous speech against the Iraq War. We had helped him create a training program for African Americans to work in campaigns, and he had been an EMILY's List keynote speaker at our lunch in 2006.

I have a young biracial godson named Jonathan, who was eight at the time, and I had always thought Barack would be a great role model for him. So, I invited Jonathan to attend our lunch and arranged for him to have his picture taken with the then senator. It occupied a prominent place of honor on my desk at EMILY's List.

On this night, however, Barack said he had to talk to me privately. I love Barack Obama. He has the most wonderful way of making you feel like he's your best friend. But I was wary—and with good reason. The next day, he called with a message that filled me with mixed feelings. He was running for president, he said, and he wanted my support.

At some point, I hoped he would be president—but certainly not now. I was excited about Hillary becoming the first woman president and determined to help her win. She still wasn't officially running, but she was already putting her campaign together. In fact, there was an important role for me. A few weeks earlier, before Christmas, she had called to ask me to be one of the chairs of her campaign.

Yes, I felt personally connected to Barack. But I had my candidate. In my mind, it was simple: first Hillary, then Barack. I laughingly told him he was breaking my heart. Then, I hung up the phone and wistfully took the photo of Barack and Jonathan off my desk.

A bit more than two weeks later, on January 20, 2007, Hillary formally announced she was launching an exploratory committee for the presidential election of 2008. Everyone had been expecting the announcement for ages. The idea of a woman becoming president had been beyond our wildest dreams when we started EMILY's List, and now Hillary was being hailed as the "inevitable" Democratic nominee.

At EMILY's List, we had already started to prepare an endorsement that would appear on our website and go out in a massive e-mail blast as soon as Hillary made it official. We were the first organization to endorse her, and we immediately began raising money for her campaign. "I am one of the millions of women who have waited all their lives to see the first woman sworn in as president of the United States, and now we have our best opportunity to see that dream fulfilled," I told reporters.

By February, nine candidates had already launched campaigns for the Democratic nomination.* Initially, Hillary's team thought John Edwards, the good-looking, smooth-talking young senator from North Carolina, would pose the biggest challenge to her. But on February 10, when Obama announced he was running, as a young, charismatic African American, he became an almost irresistible symbol of change. By the end of the first quarter of 2007,

---

* In addition to Hillary, there were Sen. Joe Biden (D-DE), Sen. Chris Dodd (D-CT), Sen. John Edwards (D-NC), former senator Mike Gravel (D-AK), Rep. Dennis Kucinich (D-OH), Sen. Barack Obama (D-IL), New Mexico governor Bill Richardson, and Iowa governor Tom Vilsack.

Obama had matched Hillary in fund-raising, with $20 million each, and Edwards was trailing with $12 million.

The more time I spent with Hillary, the more impressed I was. I had always been drawn to her because of her groundbreaking work on children's issues, women's issues, and health care. But as I saw her speak to small groups, I was amazed at her vast knowledge and experience. I listened as she gave a fifteen-minute answer on a question about Pakistan and nuclear arms that was a fascinating seminar of the nuances of Middle East policy. There wasn't an issue related to health care that she didn't have an in-depth understanding of. The more I heard, the more I believed she would be the best candidate for president.

But I was surprised to discover that not all EMILY's List members agreed with me. To some of our donors, Hillary evoked the bitter partisanship and ugly battles of the Clinton administration: Whitewater, Filegate, Travelgate, and Monica Lewinsky. Obama, however, was a fresh new face who summoned up the compelling possibility of a new kind of politics. A few asked why Hillary stayed married to Bill after the Monica Lewinsky scandal. Was she just an ambitious opportunist? Hillary was divisive, others said. She couldn't win, and she compromised on her principles. And the fact that Hillary had voted for the Iraq War resolution was a serious problem. We certainly had experienced times when members preferred another candidate in a primary. But this was on the national stage, and more members were weighing in. All of a sudden, we were having conversations more difficult and more polarizing than any we had had before. Again and again, I heard someone say, "I'm for Obama. I don't want to support EMILY's List if it's going to help Hillary." It was shocking at times.

On the campaign trail, there were other issues. Women candidates had always faced gender stereotypes. With Hillary, it was no different. By this time, she had won two senatorial elections, had an impressive catalog of legislation to her credit, and had helped revive New York after 9/11. As the first woman professional in the White

House, she had been the most activist First Lady since Eleanor Roosevelt. She had also carved out a career as one of the most important legal scholar-activists in the country and had led educational reform in Arkansas. She was the first woman to be a law partner in the Rose Law Firm.

But for some people, that wasn't enough, and the sexism often manifested itself in the most vile, hateful forms imaginable. There were Hillary Clinton nutcrackers sold as novelty items ("legs confidently spread, ready to crack some nuts with her stainless steel thighs") and hate-filled obscenities on the Internet. Babies' bibs sported the slogan I WISH HILLARY HAD MARRIED O. J. There was a Facebook page called "Hillary Clinton: Stop Running for President and Make Me a Sandwich."

And it wasn't just the right hurling epithets. On MSNBC, the supposedly liberal cable channel, Chris Matthews referred to Hillary as a "she devil," as "witchy," as Nurse Ratched and Madame Defarge. He called her male supporters "castratos in the eunuch chorus." Other pundits saw fit to have discussions about Hillary's ankles — not her foreign policy. This was political commentary? Gender bias — and rage — was alive and well. Naively, I was shocked that commentators could get away with making such grossly sexist attacks in this day and age.

Meanwhile, in contrast with the way they treated Hillary, jaded reporters, pundits, and bloggers rhapsodized about Obama. "The historians can put aside their reference material," wrote Bob Herbert in the *New York Times*. "This is new. America has never seen anything like the Barack Obama phenomenon." Liberal blogger Ezra Klein was hypnotized, asserting that Obama "is the triumph of word over flesh, over color, over despair . . . Obama is, at his best, able to call us back to our highest selves, to the place where America exists as a glittering ideal, and where we, its honored inhabitants, seem capable of achieving it, and thus of sharing in its meaning and transcendence." Talk-show icon Oprah Winfrey campaigned for Obama in messianic terms. Referring to *The Autobiography of Miss Jane Pittman,* a novel and movie from the seventies, Oprah told how the protagonist repeatedly asked children, "Are you the one? Are you the one?" "Today, we have the answer to Miss Pittman's question," Oprah said. ". . . I do believe he's the one."

This was what Hillary was up against.

I SAW IT ALL FIRSTHAND that winter of 2007–2008 when I made several trips to Iowa, the first primary state, campaigning for Hillary. Having been crowned by the media as the "inevitable" Democratic nominee, Hillary needed to win Iowa if she was to maintain the perception that she was the front-runner. But her campaign was hampered by choices made during her 2006 Senate campaign. As Joshua Green pointed out in *The Atlantic,* Hillary didn't make trips to Iowa then, because she was afraid that New York voters might "punish her at the polls" if she announced her presidential ambitions so early. Also, Bill Clinton hadn't campaigned in Iowa in 1992, because Iowa senator Tom Harkin was running for president and was clearly the favorite son. As a result, Hillary had no on-the-ground political experience there. That put her at a distinct disadvantage compared to her rivals.

I drove around the state doing small events and press for Hillary. We started in Des Moines and drove east on Interstate 80 past countless cornfields and so much farmland I joked that I had begun to recognize the cows. It was very pretty in a cold, freezing kind of way.

Whenever I saw Hillary at these events, I worried about the way she was interacting with the voters. Iowa is the opposite of major urban centers, where opinions are shaped through massive media buys. People who show up at the Iowa caucuses expect to have had face time with all the candidates. They want personal contact. But Hillary arrived at one event after another with her security team, very protected and remote, gave her talk, took questions, shook hands, then got in her car and left. She was running a perfunctory, impersonal campaign in a state where politics is intimate. It was completely wrong.

Worse, I saw what a terrible process the caucus system was— particularly for constituents who supported Hillary. Although only about 1 percent of the nation's delegates are chosen in its caucuses, Iowa generates enormous media attention nationally and tends to have a disproportionate influence, because it is the first presidential competition. But Iowa voters can't simply vote for the candidate of their choice. Instead, delegates are selected by a complex caucus process that I observed on Thursday evening of January 3, 2008. It started with about three hundred people milling around a high school gym. Finally, a man told everyone supporting Barack Obama, Hillary Clinton, John Edwards, Bill Richardson, and the other candidates to go to their respective designated areas. An area was also designated for undecided voters, and for a period of roughly thirty minutes, all the other participants were allowed to lobby the undecideds to win support for their chosen candidates.

When the lobbying ended, everyone returned to his or her designated area, and the caucus officials determined which candidates were viable, which meant they had to have at least 15 percent of the vote. There were tons of people for Obama, a lot for Hillary, and a medium-sized group for John Edwards. The Bill Richardson contingent, among others, didn't make the cut, so all of a sudden a half dozen or so of Richardson's supporters were up for grabs. Soon, people were running around like maniacs, cutting deals, cheering, and yelling, "Come with us! Come with us!" It was like Walmart during a Black Friday sale.

It's basically the highly educated, politically sophisticated, and more affluent who go to the caucuses, and this fact played to Obama's strengths. Hillary, by contrast, did extremely well with working-class Democrats, but they were not as likely to be caucus participants. All over Iowa, I met people who said they loved Hillary, but they couldn't get off work to go to the caucuses, or they had kids and couldn't afford a babysitter. At a senior retirement center, I met elderly women who were thrilled by the possibility of seeing a woman president, but it was too cold and icy for them to go out at night. And in a caucus, there is no absentee voting.

To me, this was democracy at its worst. The entire process was devoid of privacy. Just imagine two spouses who disagree about whom they support but who don't really want to get into a fight in front of all their neighbors. The caucus process was so complicated that it was not surprising that many eligible women voters didn't bother to attend. Again, that hurt Hillary — not Obama.

When the votes were counted, on January 4, Obama won with 37.6 percent, John Edwards had 29.7 percent, and Hillary came in third with 29.5 percent. It was a devastating defeat that completely changed the complexion of the race. Hillary's wonkish pragmatism had come off as uninspiring next to the message of hope and change Obama presented.

She had lost so decisively that Democratic strategist Bob Shrum told the *New York Times* that the upcoming primary in New Hampshire the next week was a question of do or die. "If Hillary doesn't stop Obama in New Hampshire, Obama is going to be the Democratic nominee," he said.

But the New Hampshire primary was coming right up, on January 8, and Obama, riding the crest of his victory in Iowa, had a seemingly insurmountable lead of 13 points in the polls.

FOR ME, IT WAS ON TO New Hampshire for Hillary. I went to a big rally at the auditorium in Manchester. Hillary was phenomenal, but the press continued to rave about Obama.

The next day, January 5, was the occasion of the presidential debate at Saint Anselm College in Goffstown, New Hampshire, and it was here that the dynamics of one of the most hotly contested primary races in history played out on TV. Four candidates made the cutoff for the debates — Hillary, Richardson, Edwards, and Obama — with the latter two ganging up on Hillary as never before, casting her as a representative of the status quo in contrast to champions of the profound transformations they hoped to bring about in the body politic.

Hillary fought back defiantly, pointing out that Obama's eloquent rhetoric did not change the fundamental hard, cold political realities of Washington. "Making change is not about what you believe," she said. "It's not about a speech you make. It is about working hard . . . We don't need to be raising the false hopes of our country about what can be delivered."

Finally, Hillary was asked why voters didn't find her as likable as Obama. "Well, that hurts my feelings," she said. As she spoke, she smiled wanly, as if somewhat wounded. The entire audience laughed and applauded. "But I'll try to go on," Hillary added. Her smile widened with a certain mock stoicism, and everyone laughed.

"He's very likable," Hillary continued. "I agree with that. I don't think I'm that bad."

All over the country, Hillary was winning sympathy, and Obama had to cut her off. "You're likable enough, Hillary," he said. He spoke as if he were grudgingly conceding that his opponent had scored some points.

As I watched, I recoiled. I thought to myself, "Women are going to watch this and go crazy." My belief in Hillary was stronger than ever.

Lots of women shared my feelings. The media had fallen head over heels in love with Obama, but the whole thing—Edwards, Richardson, Obama, and moderator Charlie Gibson ganging up on Hillary—was like watching the boys in the locker room engaging in their male-bonding rituals and completely dismissing an extraordinarily accomplished woman.

AS I TRAVELED AROUND New Hampshire, I met scores of young men and women who had worked for EMILY's List or participated in our training programs. Whatever Hillary's travails, one of the few reliably upbeat elements of this campaign was going into one small town after another and finding young people working for Hillary who would say, "I was in your training program, and it helped me get a job in a state race." Meanwhile, time had almost run out for the January 8 primary. The day before voting, Hillary explained to an undecided woman voter exactly why she went into public service. "I couldn't do it if I just didn't passionately believe it was the right thing to do," she said. "I have so many opportunities from this country, I just don't want to see us fall backwards. This is very personal for me — it's not just political, it's not just public. I see what's happening. We have to reverse it. And some people think elections are a game, they think it's like who's up or who's down. It's about our country. It's about our kids' futures."

Hillary didn't cry, but her voice quavered and she was visibly moved. Everyone had been calling her cold and calculating, but you could see that what really moved her was the idea that she might make a difference for people.

Not that the press saw it that way. "Can Hillary cry her way back to the White House?" asked Maureen Dowd in a column in the *New York Times,* asserting that Hillary projected a "whiff of Nixonian self-pity." Nixonian? I sure didn't think so. Dowd also quoted a man who said the episode raised the question of how Hillary would behave as a woman president during wartime. "Is this how she'll talk to Kim Jong-il?" he asked.

Dowd cited others who dismissed Hillary's display of emotion as nothing more than a calculated act that raised questions about her authenticity. "That crying really seemed genuine," one reporter jokingly told her. "I'll bet she spent hours thinking about it beforehand." And this was the *New York Times* — not Fox News.

Not to be outdone by the press, hecklers yelled, "Iron my shirt!" at one of Hillary's last appearances in New Hampshire.

"It hurt," Hillary told me later. "It was every emotion you could imagine. I don't have any illusions, but explicit sexism was a surprise to me."

★　★　★

BUT THERE WAS ALSO a brighter side to these last few days in New Hampshire. I went to Manchester and other towns near the coast of New Hampshire, and there were HILLARY signs everywhere. Often, I saw couples walk by Hillary's campaign signs, and the woman would quietly give a thumbs-up. And at these last rallies before the voting, Hillary stood there answering questions, personable, connected. I realized she had taken control of a campaign that had huge problems, and it was impressive to watch her take charge. Hillary was a fighter.

My friends from EMILY's List were fighters, too. With Hillary's back against the wall, the Majority Council Development Team flew to New Hampshire to help. We had sent a crew of young staffers and interns to New Hampshire to join the get-out-the-vote effort. I would meet the staffers for a short stint of visibility on a town square, then head off to the next town.

One of the women later told me of a chance encounter she had with a young African American man. "This young guy came up to me and said, 'Do you realize what it means to me to be voting for an African American president?'

"'Yes, I do,'" she said. "'Because I think you might know what it means to me to be voting for a woman. And I'm just sorry they both can't win.'

"'Fair enough,' he said."

And at that moment, I thought, *My God, what an election. But one of them is going to lose.*

When I called my friends in Washington that day, I knew pundits were talking about the Obama "tsunami" and saying he would win by as much as 20 points. But I wasn't upset. Our chances felt good to me.

On Election Day in New Hampshire, I got up early and went to the polls for Hillary. I worked at the polls during the day with Dan Schwerin, one of Hillary's young aides, and drove to Manchester for the election-night party. When the polls finally closed, Dan got a call from a town near the coast.

"We won that town," he said. "And that's not good territory for Hillary."

Then, someone else called with a similar message. Then, a third.

"You don't think we can pull this off, do you?" I asked.

More good news came rolling in.

By the time we finally got to the convention center in Manchester, the place was electric. I went to the VIP section and was standing on the floor with Los Angeles mayor Antonio Villaraigosa. Suddenly, CNN projected that Hillary had won. There was pandemonium.

Hillary had won New Hampshire! The polls were completely wrong. The pundits were completely wrong. Best of all, she was back in the race. Women and independents had turned the tide. For all the snarky coverage of Hillary, her "human" side seemed to have won her votes. She had carried the women's vote decisively.

Finally, after a hastily conceived victory speech, Hillary made her way to the rope line to shake hands. When she got to me, she put her hand on my arm. She had a special message.

"Ellen," she said. "I'm the first woman to ever win a presidential primary."

It was such a great night.

OF COURSE, HILLARY'S New Hampshire victory didn't mean she was the front-runner again. Quite the contrary. Obama was still being hailed as someone capable of opening a new era of "postpartisan" politics that would transcend the tawdry realities of Capitol Hill. And so, over the next four months, Hillary and Obama waged the most dramatic neck-and-neck primary battle for the Democratic nomination in recent history. Many expected it to be settled once and for all on Super Tuesday, February 5, during which no fewer than twenty-three states and territories held primary elections or caucuses. But even that was a virtual tie.

In March, Hillary won Ohio and Rhode Island convincingly, and scored a narrow victory in Texas. But Obama still had a lead in pledged delegates. Even in Texas, where Hillary won the popular vote, her campaign failed to master the complicated delegate-selection process, so Obama ended up with more delegates than she.

The nominating process was pointing out a fatal flaw in the Clinton campaign. The nomination was determined by who had the most delegates, not who had the most votes. In state after state, Obama would collect more delegates even as Clinton racked up the votes. It was as if the Clinton campaign had never read the rulebook. It was absurd and disheartening. As the campaign went on, Hillary seemed to get stronger and stronger as a candidate, yet Obama was moving closer and closer to victory.

On April 22, Hillary won a substantial victory in Pennsylvania, a key swing state that was crucial to Democratic hopes in November. But on May 6, Obama won North Carolina by 15 points. Obama began to act as if he had the nomination wrapped up and was most concerned about the general election in November against presumptive Republican nominee John McCain.

But Hillary wouldn't give up. She won convincingly in West Virginia and Kentucky. Then, Obama took Oregon, and for the first time had a solid majority of pledged delegates.

On June 3, the day when the last two primaries were held, in South Dakota and Montana, about sixty superdelegates endorsed Obama. That meant he was now over the "magic number" of 2,117 delegate votes necessary to win the nomination. Later that day, the media declared that Barack Obama was the presumptive nominee. We were all devastated. Hillary had no choice but to concede and endorse Obama.

In the end, Hillary had staged an astounding comeback. When the primaries were over, if one counted Michigan's highly controversial primary,* she had actually beaten Obama in the popular vote — 17,857,501 to 17,584,692 — and she did especially well by winning

---

* The Democratic National Committee determined that Michigan violated the party rules by changing the date of its primary, and refused to seat its delegates at a convention. Because of the controversy, Barack Obama did not put his name on the ballot in Michigan.

vital swing states such as Ohio, Pennsylvania, and Florida. But that wasn't enough, because her campaign repeatedly failed to master the procedures by which delegates were allocated.

Then, on Saturday June 7, 2008, Hillary brought thousands of her supporters together in the National Building Museum in Washington for one last time. I took my seat in the VIP section, noticing that there was no one sitting in front of me. But just before Hillary took center stage, Bill and Chelsea Clinton appeared and sat there.

Then, Hillary Rodham Clinton proceeded to give the most dramatic and commanding speech of her life. Her delivery was stronger and more forceful than ever. If there had ever been questions about what her voice was, this was it. "You can be so proud that, from now on," she told her supporters and staff members, "it will be unremarkable for a woman to win primary state victories, unremarkable to have a woman in a close race to be our nominee, unremarkable to think that a woman can be the president of the United States." She said, "To those who are disappointed that we couldn't go all of the way, especially the young people who put so much into this campaign, it would break my heart if, in falling short of my goal, I in any way discouraged any of you from pursuing yours.

"As we gather here today, . . . the fiftieth woman to leave this earth is orbiting overhead," she continued. "If we can blast fifty women into space, we will someday launch a woman into the White House . . . Although we weren't able to shatter that highest, hardest glass ceiling this time, thanks to you, it's got about eighteen million cracks in it."

It was devastating. For twenty-nine minutes, she held thousands of us spellbound. Hundreds of people were weeping, and it took everything I had to keep from breaking into tears. But because about fifty photographers were kneeling and taking photographs of Bill and Chelsea, I was in their line of fire, and the last thing I wanted was a picture of the president of EMILY's List behind Bill and Chelsea, bawling away. My eyes were dry, but my heart was broken.

Seventeen

EMILY'S LIST 2.0

TERRIFIC AS HILLARY'S SPEECH was, it didn't change the fact that this was still an extraordinarily sensitive time for us at EMILY's List. A lot of Hillary supporters still had deeply mixed feelings about the campaign outcome — and I was one of them. "I can still be disappointed and also determined to elect Barack Obama," I told one reporter. "They're not mutually exclusive."

Now that the primaries were over, EMILY's List executive director Ellen Moran and I agreed to put aside our disappointment and do everything possible to help Obama win the general election in November. On Tuesday, June 10, we went to Chicago to meet with Obama campaign manager David Plouffe. We wanted to make it clear that we were determined to get out the women's vote for Obama in critical states. Specifically, we told Plouffe we were going to run massive WOMEN VOTE! programs in New Hampshire, where Jeanne Shaheen was running for Senate, and in North Carolina, where Kay Hagan and Bev Perdue were running for the Senate and governorship, respectively. Women and younger voters could help put Obama over the top.

I did everything I could for Obama — even when it meant swallowing my pride. No event was more difficult for me than the National Partnership for Women and Families 2008 Annual Luncheon in Chicago, on June 20, at which, as chair of the organization's board, I was scheduled to speak along with Michelle Obama. I first met Michelle in 2002, when Barack was a state senator. I knew the Obama campaign wanted a picture of me with Michelle, but I was still hurting from Hillary's loss and feeling petulant. Then, as I watched Michelle speak, I thought, "Ellen, get over it. This woman is working her heart out for her husband. She doesn't like politics, but she's doing all the things that need to be done. Give her a break."

So, when Michelle finished, I stepped forward, gave her a big hug, and all the photographers snapped away. Sometimes in politics, as in life, you just have to swallow hard and do the right thing.

The postprimary challenges continued, especially for the EMILY's List fund-raising team. Obama supporters were angry that we had backed Hillary. And do-or-die Hillary loyalists were mad because we now supported Obama. Some were even upset that we hadn't done enough for Hillary, which was the oddest criticism of all. We all tried to put on a good front, but it was agonizing. We reminded members about the other important races we were involved with and the importance of WOMEN VOTE! for both Obama and our women candidates. Slowly, the anger dissipated, and many members came back. Some members never have.

At the Democratic National Convention in Denver, in August, EMILY's List had a big unity event for women, with Nancy Pelosi, Michelle Obama, and Hillary all speaking. It was an important moment for Democrats when Hillary, Michelle, and I walked in together — a critical statement that Democratic women would be unified in working to elect Barack Obama and all the Democrats on the ticket.

MEANWHILE, SOMETHING ELSE was happening that profoundly affected the upcoming election. After years of virtually unregulated trading of subprime mortgages, credit swaps, and other arcane financial instruments, the financial world was starting to implode. On September 15, Lehman Brothers filed for the largest bankruptcy in U.S. history. In rapid succession, other institutions, such as Fannie Mae, Freddie Mac, Merrill Lynch, Citigroup, and AIG failed, were acquired under pressure, or became subject to government takeover. Stocks, housing prices, and retirement assets plummeted.

The only bright side of the crisis was political. Taking place just weeks before the November elections, it left the Republicans more vulnerable than ever. All of this had occurred on their watch.

As a result, even with the disappointment of Hillary's defeat, the 2008 election was historic for us. We put together a big WOMEN VOTE! campaign in New Hampshire and North Carolina that was very important to Obama — and to electing two new women senators, Jeanne Shaheen and Kay Hagan, in New Hampshire and North Carolina, respectively, and Gov. Bev Perdue, also in North Carolina.

Altogether, we elected twelve new women to the House of Representatives. For the fall campaign, we had raised a total of $43 million, by far our highest total. Plus, of course, after a very serious bid to put the first woman in the White House, we ended up helping the Democrats sweep the White House and both houses of Congress. After a difficult year, victory felt especially sweet.

NOT LONG AFTER the election, we also got an unexpected bonus, thanks largely to the efforts of our POP program and the new relationships we were building in the states. It came about in December, when President-elect Obama nominated Hillary as secretary of state, an appointment that helped heal the Obama-Clinton breach in the party and, in the process, created an open seat from New York in the Senate.

As always, EMILY's List had its eye on open seats, and in this case, we were well positioned, thanks to the effective partnership between our POP program and Gov. David Paterson that we had initiated when he was New York state senate minority leader. Now that he was governor, Paterson had the authority to appoint Hillary's successor — and he was more than receptive to the notion of appointing a woman. Among the possibilities I discussed with him were Reps. Carolyn Maloney, Nita Lowey, and Louise Slaughter. I also told him that Kirsten Gillibrand was a tenacious fund-raiser and a terrific campaigner who had done very, very well in winning what had been a Republican congressional district. After all, whoever the governor appointed would have to run for the full term at the next election. Having a strong candidate was crucial.

But before the decision process had gotten anywhere, a new and unexpected name came into play: Caroline Kennedy announced she was interested in the seat. As an author, attorney, and daughter of President John F. Kennedy and First Lady Jacqueline Kennedy, Caroline was an iconic figure who immediately won support from Rep. Louise Slaughter, New York mayor Mike Bloomberg, former mayor Ed Koch, and others. But beyond her family background, she had almost no real political experience and soon came under fire for not having voted in various Democratic primaries and general elections since registering, in 1988 in New York City, among other things. On January 22, Kennedy announced she was withdrawing her name from consideration. After another conversation with the governor, I was delighted when he reported that Kirsten Gillibrand would be the next senator from New York.

MEANWHILE, ON JANUARY 20, 2009, Barack Obama had been sworn in as the first African American president of the United States. Much as I would have loved to see Hillary there instead, it was a truly historic and exciting event. In addition, I got what seemed like a replay of Bill Clinton's first days in office when, on January 29, I was invited to the White House to watch Obama sign the Lilly Ledbetter Fair Pay Act, which expands the Civil Rights Act of 1964 to make it easier for employees to sue for wage discrimination. Two of the bill's leading sponsors, Sen. Barbara Mikulski and Rep. Rosa DeLauro, stood with Obama as he signed this bill into law, flanked by Ledbetter, Hillary, House Speaker Nancy Pelosi, and others who worked for its passage. This image was in striking counterpoint to a 2003 photo I saw of President Bush signing the so-called partial-birth abortion ban surrounded by men in dark suits.

With Barack Obama in the White House, we had indeed entered a new era, and I was beginning to think it was time for a new era at EMILY's List as well. For one thing, after nearly twenty-five years, I was concerned that I was going stale. For twenty-five years, I had developed a vision of how we could elect women to office. I was proud of my ability to translate our political strategies into compelling messages that convinced members to contribute. But I was beginning to repeat myself. If I was bothered by my repetition, surely the members would begin to be bothered as well.

Our major donors had been thirty-five to sixty years old when EMILY's List began. Now, those same donors were sixty to eighty-five. As I reached my mid-sixties, I was confident that I knew how to speak the language of those members, but my voice didn't seem right for attracting younger members.

I also wasn't interested in learning the new, Internet-based forms of communication. Personally, I was much too private to spend time posting on Facebook. Tweeting in 140 characters seemed silly. With horror, I realized I had become — shudder — old-fashioned. As a child of the sixties, I was appalled at myself.

I knew that EMILY's List had to master the Internet and the world of social media. Because I had little affinity for it, I began to consider hiring a new, younger leader. It was definitely time for a redo.

We needed EMILY's List 2.0. And that called for someone who could be a leader on both the political front and the marketing front. I knew, of course, that there were wonderful fund-raisers who didn't know the first thing about winning political campaigns, and there were brilliant political strategists who couldn't raise a dime. I was looking for someone who had both sets of skills. That was going to be an incredible challenge. So, I hired a headhunter and began to make calls. Meanwhile, our former political director, Martha McKenna, suggested Stephanie Schriock.

Stephanie was one of the hottest properties in American politics. In 2004, as national finance director of Howard Dean's presidential campaign, she helped revolutionize political fund-raising by using the Internet to help raise $52 million for the Democratic primaries. Then, in 2006, she returned to her home state of Montana to be campaign manager for Jon Tester, who was looking to dethrone three-term Republican incumbent Conrad Burns in a state with an 8-point Republican advantage. Tester won by the thinnest of margins — a mere 3,562 votes — and was so impressed by Stephanie that he made her his chief of staff.

In 2008, Stephanie left Tester temporarily to work as campaign manager for Al Franken, the veteran *Saturday Night Live* comedian, who was trying to unseat Sen. Norm Coleman, a well-funded Republican incumbent. Tester's "loan" of Stephanie proved to be extraordinarily generous, as she worked day and night up until Election Day — and then an additional eight months through bitterly contested recounts. This contest made her Montana squeaker look like a landslide. Al finally emerged victorious by 312 votes out of roughly 3 million, in one of the closest elections in American history.

Stephanie returned to her work with Senator Tester. A few months later, in the fall of 2009, Martha McKenna called to see whether she was interested in leading EMILY's List. "I had no intention of leaving [Tester] after his immense patience," said Stephanie. "He was up for reelection, and I thought I had three years to focus on getting him reelected. I was also concerned that Ellen's shoes were impossible to fill and that the board, staff, and members wouldn't accept someone new after twenty-five years. But I was really intrigued by the job. I do love tough gigs — hell, I worked on the Franken campaign!"

This also happened to be the period during which Congress debated President Obama's health-care package, which raised issues about access to reproductive care. "I was working for great men who I still support," said Stephanie. "But I knew I was working for an institution [the Senate] in which there clearly were not enough women, and I knew how important it was that the voices of women be heard."

After Stephanie met twice with our search committee, I offered her the job. She started in February 2010, and I stayed on as chair of the board.

All this wasn't easy for me. Changing leaders is difficult in any institution, but it is particularly difficult when a founder of the institution tries to give up leadership. It was hard emotionally to let go of what I had created. I wanted EMILY's List to change, but I didn't want anything to be different. I wanted a new leader, but I didn't want to let go. After all, this was my baby. It took me awhile to realize how absurd my thinking was.

My friends almost universally told me the transition wouldn't work. I quickly tired of stories about founders who could never let go. Stephanie had her own challenges, learning how to be the public voice of EMILY's List and dealing for the first time with a board of directors.

But both Stephanie and I were determined not to fail. The transition took a lot of work, and, frankly, we needed a number of tough conversations with each other, a consultant, and a good eighteen months before we figured it out in a way that was comfortable for both of us. In the end, we emerged in a very positive place, and EMILY's List is stronger as a result. I know one of the reasons I was able to let go is largely that I could see what a terrific job Stephanie was doing. I had dreamed that EMILY's List would continue to grow and elect even more women. When I became convinced that was happening, I could relax in my new role.

STEPHANIE'S FIRST PRIORITY was expanding our online presence. For many years, the only way a person could be considered a member was by giving money to candidates and/or to EMILY's List. But now we redefined membership to include people who helped in many additional ways—by signing an online petition, forwarding content to friends, and encouraging friends to vote (which happens to be an enormously effective mechanism). "EMILY's List had this amazing foundation of deeply committed supporters and had been unbelievably successful strategically," said Stephanie. "But a lot of people still didn't know about it."

So, EMILY's List invested more than a million dollars in redoing the entire database and social-media program. We began updating our content regularly and expanding our online presence. We built massive e-mail lists with an eye toward creating a community of people who interacted with EMILY's List and who, over time, would become donors.

Soon enough, using our virtual Rolodex, visitors were able to upload their contacts and invite them to join the EMILY's List community. We encouraged readers to comment on our Facebook page, watch candidate ads, and pledge to vote in November. Our online community, which had been only 127,000 in 2008, soared to nearly 500,000 in 2010. It was a great new marketing tool for us. It's challenging to convert people from signing a petition online to becoming a donor, but if you have a huge database, you can instantly test everything at almost no cost.

INITIALLY, THE REPUBLICANS didn't know how to handle Obama and his overwhelming popularity, but soon enough, Rush Limbaugh, the bombastic radio talk-show host who had become a powerful voice in the Republican Party, started asserting that the Republicans were nothing but a bunch of wimps and had to fight back.

The first signs of a new, energized political force became apparent to me in Massachusetts in late 2009. Sen. Ted Kennedy had died, and a special election was scheduled for January 2010. Initially, Martha Coakley seemed a shoo-in to win Kennedy's seat, in part because she was a popular Democratic state attorney general and in part because the seat she was running for had such a rich Democratic heritage. John F. Kennedy had held it as senator, and, after he was elected president, his brother Teddy filled it for the next forty-seven years. In addition, Coakley's opponent, state senator Scott Brown, was thought to be something of an empty suit. Having been named America's Sexiest Man in 1982 by *Cosmopolitan* magazine (complete with a seminude centerfold), he drove around photogenically in a pickup truck while his daughter won attention for her appearances on the TV talent show *American Idol.*

But Coakley turned out to be a problematic candidate, for example, repeatedly making comments scorning the Boston Red Sox, a capital offense in Massachusetts. The very skills that made her a good attorney general — delivering a well-reasoned argument — made her appear wooden and unapproachable as a candidate. At the same time, thanks to the newly formed Tea Party, the Obama administration's health-care plan ignited an enormous right-wing furor. Even though the national Tea Party did not endorse Brown outright, affiliated groups such as the Greater Boston Tea Party organized a fund-raising breakfast for Brown, and the Tea Party Express took out ads supporting him on national cable TV. In the end, Brown won, 51.9 percent to 47.1 percent, a major upset in a hugely Democratic state.

The fact that Teddy Kennedy's seat had gone to Brown was an alarming wake-up call for old-line Republicans, in that it signaled the emergence of the Tea Party as a powerful new force on the right, one the GOP establishment did not control. During the next nine months, Tea Party–backed candidates such as Sen. Rand Paul (R-KY), Sen. Mike Lee (R-UT), Christine O'Donnell of Delaware, and Sharron Angle of Nevada defeated Republican establishment candidates, including incumbent senators, former governors, and the like, in a seismic shift for the party. The Republican establishment was losing its grip. This was the Republican equivalent of a Carol Moseley Braun moment, in the sense that candidates who had previously been on the fringes were now electable. This time, however, extreme conservatives were in the driver's seat. The Republican Party had no choice but to become more conservative and appease the right. There were roughly two hundred Republicans in rock-solid Republican seats who didn't have the slightest worry about losing to Democrats in the general elections. As a result, they were far more afraid of what Rush Limbaugh might say about them, and they tacked far to the right. This was yet another step toward the extreme partisanship we see today.

On the bright side, in November, we elected Terri Sewell as the first African American congresswoman from Alabama, along with four more new-to-Congress women of color. And, in the Senate, Barbara Boxer, Kirsten Gillibrand, Patty Murray, and Barbara Mikulski all won reelection. But in the end, 2010 turned out to be the worst election cycle since 1994. Democrats lost a staggering sixty-three seats in the House, which meant the Republicans took over and Nancy Pelosi had to step down as Speaker.

NEEDLESS TO SAY, WE WERE determined to reverse things in 2012. One of the most exciting candidates Stephanie recruited was Harvard Law School professor Elizabeth Warren, who, in the wake of the 2008 fiscal crisis, had chaired the congressional oversight panel that evaluated the government bailout and related programs.

A legal scholar who was both nonpartisan and apolitical, Warren first heard of EMILY's List when she was at the University of Texas, in the eighties. "I was a teacher and a young mother back in Austin, Texas, when I got a letter in the mail asking me to support women running for office," she said. "I believe I wrote a check for twenty-five dollars." She had never made a political contribution before.

More than twenty years later, as a tenured professor at Harvard Law School, after doing an enormous amount of research on how the American middle class was crumbling, Elizabeth had become one of the widely cited law professors in the United States and was asked to chair the congressional oversight panel to put some accountability into the government bailout and related programs that helped resolve the 2008 fiscal crisis.

Not long afterward, in September 2010, President Obama appointed Warren assistant to the President and special adviser to the secretary of the treasury to help set up the U.S. Consumer Financial Protection Bureau. Elizabeth was the obvious choice to head the new agency, because it was her idea, but Senate Republicans saw her as such a zealous consumer advocate that President Obama backed off giving her the appointment. So, it was time for her to head back to Massachusetts.

Meanwhile, EMILY's List was looking for senatorial take-back seats in 2012, and by May 2011 Elizabeth's name was already being circulated in the press as a potential contender. "I said with complete honesty that was nowhere on my list," Elizabeth told me. "I wasn't thinking about it at all."

But she had not met Stephanie Schriock.

In July 2011, as Elizabeth and her husband, Bruce Mann, also a Harvard law professor, were preparing to move back to Massachusetts, Stephanie called to see whether she wanted to run for the Senate. Elizabeth said no, but Stephanie insisted on coming over.

When Stephanie arrived, Elizabeth was in shorts, and Bruce was packing boxes for the move. As they sipped iced tea, the couple's big golden retriever slobbered all over Stephanie as she made an hourlong pitch about why Elizabeth should run, what a campaign really looked like, and how she would win.

"I wanted to tell her how it works," said Stephanie, "because if you don't already know something about running for office, it looks like an impenetrable barrier." So, she went into the nuts and bolts of running a campaign and ended by telling Elizabeth that if she ran, EMILY's List would be with her every step of the way. "That changed something that looked completely impossible, something that one couldn't even think about, and made it something where you could see that there was a path," said Elizabeth.

Elizabeth still wasn't completely convinced, but at least Stephanie had opened the door. Then, not long afterward, Sen. Patty Murray called Elizabeth. "I immediately started out with my long list of why I'm not prepared to be senator," said Elizabeth. "About why I'm not qualified right now. I said, I haven't learned about this and I haven't learned about that."

Finally, Patty interrupted with a laugh. "You're such a girl," she said. "Please. Men never ask if they have all the right experience. They just ask if you can raise enough money for them to win. That's it. That's all they ever want to know. It's always women I have this conversation with. Women, who, in effect, feel they have to be perfect to do this, that they need to know everything about every committee they'll be working on, that they already need to be an expert on every issue, or otherwise they better not run."

"So, there were still lots of things I had to think about," Elizabeth said, "but this was no longer one of them. Patty just told me, 'You work hard, you're smart, you'll figure it out.'" Elsewhere, in Wisconsin, in early 2011, four-term senator Herb Kohl (D-WI) had let it be known he was not going to run for reelection. That meant we had an opening for a woman, and after four terms in the House, Tammy Baldwin was more than ready.

But first, Tammy had to clear the decks within her own party. Former senator Russ Feingold had been defeated in 2010 and was considering running for the seat. "He was beloved by many in Wisconsin, and there were certainly any number of people who were urging him to consider it," Tammy told me.

When Feingold told her he was thinking about it but wasn't ready to decide, Tammy had no choice but to act. "I said, 'You can wait. But I'm not known statewide, so I need to start organizing now.'" Which is exactly what she did — and, in the process, potential rivals left the field. First, Feingold declined to run, then Rep. Ron Kind, a moderate Wisconsin Democrat who was seen as a strong contender, bowed out as well. So, Tammy took the Democratic senatorial nomination unopposed while Republicans slugged it out in a bitter primary. The Republican primary winner was Tommy Thompson, the former governor and secretary of health and human services.

OF COURSE, 2012 WAS another of those supposedly magical "2" years in which we potentially could benefit from a larger-than-usual number of open seats. But, with the Republicans having taken over redistricting by seizing control of many statehouses and state legislatures in 2010, they controlled the process.

We gained traction in another way, though. There were now so many hard-line right-wingers among the Republicans that they launched what amounted to a war on women, and that provided a wake-up call to our constituency. By this time, there was a huge group of women in their forties and fifties, as well as younger women, who did not identify as feminists and had always taken abortion and birth control for granted.

Now, all of a sudden, reproductive rights were in jeopardy, thanks to a variety of bizarre, antiquated notions being put forth by a new breed of right-wing Republicans. When Republican representatives convened hearings on contraception mandates, they managed to present a panel with no women on it whatsoever. Then, in an effort to make sure women were heard, in February 2012, the House Democratic Steering and Policy Committee asked Georgetown University law student Sandra Fluke to testify about why university health insurance should include birth control. In response, Rush Limbaugh labeled Fluke a "slut" and a "prostitute."

Meanwhile, Rep. Cliff Stearns (R-FL), chairman of the Energy and Commerce Subcommittee on Oversight and Investigations, had launched a sweeping investigation of Planned Parenthood Federation of America, the highly regarded nonprofit that was the largest U.S. provider of reproductive health services, including cancer screening, HIV screening and counseling, contraception, and abortion. Needless to say, it was abortion that was of concern to Stearns and other Republicans, who proceeded to launch a sweeping investigation that requested internal audits from Planned Parenthood dating back twelve years, and state audits for the past twenty years for both the national organization and all of its eighty-three affiliates. Now that Planned Parenthood was under investigation, Susan G. Komen for the Cure, a large breast cancer research and advocacy organization, influenced by recent hire Karen Handel, a Republican foe of abortion, withdrew funding from Planned Parenthood, citing the Stearns investigation. The uproar was immediate. About one in four women in America has used Planned Parenthood services at one point or another. Many had walked in Komen fund-raisers to find a cure for breast cancer, and felt betrayed both by the attack on Planned Parenthood and by Komen's waging a seemingly political war. A firestorm of anger hit Komen, and donors withdrew their support. (Komen later reversed its policy, and Handel resigned from the organization.) Most famously — or perhaps infamously — Missouri Republican senatorial candidate Rep. Todd Akin, who was running against Democrat Claire McCaskill, asserted that victims of "legitimate rape" rarely get pregnant, because "the female body has ways to try to shut that whole thing down."

The clear implication was that, if women did get pregnant after rape, they actually wanted it — in other words, that women were claiming they had been raped as a way to get an abortion. It was horrifying what these men thought.

All over the country and at all political levels, Republicans challenged reproductive rights, employment rights of women, and gay rights. Former senator Rick Santorum (R-PA), a presidential candidate, insisted that birth control was "not okay. It's a license to do things in a sexual realm that is counter to how things are supposed to be." Mitt Romney, the eventual Republican presidential nominee, said that states might have the right to ban contraception. Janice Daniels, the Republican mayor of Troy, Michigan, declared she was going to throw away her I LOVE NEW YORK tote bag, "now that queers can get married there." Jay Townsend, a spokesman for Rep. Nan Hayworth (R-NY), recommended hurling "acid at those female Democratic senators" who supported the Lilly Ledbetter Fair Pay Act.

ALL THESE VILE PRONOUNCEMENTS meant that there was no shortage of ammunition for EMILY's List. We could point out that the Republican congressional committee holding hearings on *our* reproductive rights did so with an all-male panel. We could point out the danger of Todd Akin getting into the Senate, where his ludicrous ideas about women's bodies might actually become policy. We could point out that in the fifty states, state legislators introduced nearly 1,100 measures combined to restrict access to reproductive care. In other words, the hard-won reproductive rights that women had had for more than a generation were now in real jeopardy.

All of which helped wake up a new generation of women. Our POP program was in full swing, training more than 1,300 women at thirty-seven seminars in seventeen states. We raised nearly $11 million for WOMEN VOTE! and sent out 4.7 million pieces of targeted mail, aired twenty unique TV ads, made more than half a million phone calls, and delivered 148 million online impressions to targeted voters. Our membership grew by 30 percent over the previous cycle, to 165,000 members, and our online community nearly quadrupled, to 2 million members.

The result, as Stephanie put it, was "an amazing, history-making, awe-inspiring, glass ceiling–shattering election." In Massachusetts, Elizabeth Warren had encountered the same problem we had been confronting for decades: the old boys in the Democratic Party in Massachusetts thought she was a weak candidate. "I don't think we can beat Scott Brown with a Harvard professor," said Scott Ferson, who was Sen. Edward Kennedy's press secretary in the 1990s.

But EMILY's List turned out in full force for Warren, with members contributing $1.2 million to her campaign and with WOMEN VOTE! reaching out to 120,000 carefully targeted pro-choice swing voters. In the end, she defeated Scott Brown with 53.7 percent of the vote, and became the first woman senator from Massachusetts.

At the same time, in Wisconsin, Tammy Baldwin had what initially appeared to be a tough battle against Tommy Thompson, a four-term governor and one-time presidential candidate who had held a cabinet post and was as wired into Washington as they come. But our members turned out for her in force as well, also contributing $1.2 million to Tammy.

Because she had been unopposed in the primary, Tammy got a chance to define herself as a candidate while Thompson had to fight to win the Republican primary. "I put out a very basic economic message of wanting to help people get ahead again," Tammy said. "And we truthfully reminded people that since he left public service, Thompson had cashed in on all of his connections. He bragged openly about having a net worth of thirteen million dollars, about being on twenty-two boards of directors, and a half-time partner at Akin Gump [the white-shoe Washington law firm]."

Tammy didn't criticize Thompson for making money so much as point out that he was using his political connections for his own gain. So, when they debated, Thompson came off as someone who really didn't have the interests of Wisconsinites at heart and as an angry man who felt he wasn't getting the respect he deserved.

As we got closer to the fall, most polls showed Thompson ahead, but on Election Day, Tammy beat him 51.5 percent to 45.9 percent, thanks to a huge margin in the Milwaukee area and Madison. The first out lesbian had been elected to the United States Senate.

And that was just one of the historic milestones. Yet another was the election of Mazie Hirono (D-HI), who became the first Asian American woman elected to the United States Senate.

Mazie had an extraordinary story. Although her mother was an American citizen, Mazie was born in Japan in 1947. Then, nearly eight years later, Mazie's mother left her abusive, alcoholic husband and fled to Hawaii, where Mazie, her brother, her uncle, and her mother all shared a thirty-five-dollar-a-month room in a boarding house. Money was so tight that her mother had to break into Mazie's piggy bank to get money for food.

After high school, Mazie attended the University of Hawaii, got a law degree from Georgetown University, and returned to Hawaii to practice law. In 1980, she was elected to the Hawaii House of Representatives, where she passed 120 laws in the fourteen years she served before running for statewide office. After serving as lieutenant governor of Hawaii, Mazie ran for the U.S. House from Hawaii, with EMILY's List's help, and won three consecutive terms before vying for the Senate. In that contest, she won in a landslide, beating Republican Linda Lingle by 25 percent.

For the first time, there were twenty women in the United States Senate—sixteen of them Democrats. We helped the Democrats hold on to the Senate by winning nine out of ten races—an incredible record. In addition to reelecting six incumbents, we sent in three terrific newcomers: Elizabeth Warren, Tammy Baldwin, and Mazie Hirono.

We also added nineteen new women to the House—in the process picking up nine new Democratic seats. In New Hampshire, we elected Maggie Hassan as the nation's only pro-choice Democratic woman governor, and sent Carol Shea-Porter and Ann McLane Kuster to the House as part of New Hampshire's all-female congressional delegation—the first in the country. And we defeated some of the worst of the worst. Sen. Claire McCaskill beat Todd Akin. In the House, Tammy Duckworth, an Iraq War veteran who had lost both legs and partial use of one of her arms in combat, beat Joe Walsh, a Tea Party–backed representative from Illinois who had waged an angry and bitter race. And Rep. Frank Guinta (R-NH), who accepted illegal loans to his campaign of more than $300,000 from his parents, lost to Carol Shea-Porter.

It was a great 2012 election, and EMILY's List 2.0 was truly on a roll.

Eighteen

WOMEN MAKE IT WORK

AS SOON AS THE 113th Congress got under way, in January 2013, the new women legislators we helped elect went to work. In the Senate, Elizabeth Warren took on the big banks. Tammy Baldwin joined the fight for pay equity. In the House, Kyrsten Sinema (D-AZ) introduced a bill to spur domestic production of energy technology. Cheri Bustos (D-IL) introduced legislation that would prevent student-loan interest rates from doubling. Likewise, Louise Slaughter (D-NY) and Gwen Moore (D-WI) led the battle to reauthorize the Violence Against Women Act, which included protections for Native Americans, undocumented immigrants, and LGBT people. Even local women who had been elected with the help of our POP program won national attention, as Texas state senator Wendy Davis did in June 2013, when she courageously donned her soon-to-become-famous pink sneakers and back brace, and staged an eleven-hour filibuster on the floor of the Texas Senate against extreme anti-choice legislation. Wendy had previously sponsored bills dealing with cancer prevention, protection for sexual-assault victims, and government transparency.

Yet the Republican Party continued moving further and further to the right. In its efforts to defund the Patient Protection and Affordable Care Act — better known as Obamacare — the Republicans resorted to Gingrich-era tactics and once again shut down the government, on October 1, 2013. For more than two weeks, roughly 800,000 government employees were furloughed, and another 1.3 million were required to work without knowing when they would be paid.

Then, on October 17, as the shutdown moved into its third week, came a series of events that showed tens of millions of people all over the country the merits of electing women — both Republican and Democratic — to higher offices. Americans saw that even in the midst of a period of horrid partisanship that left our government in gridlock, the collaborative power of women could put our country back on track.

The Washington social scene of this era reflected the partisan split that divided Congress. Gone were the days of bipartisan dinners and friendships that crossed party lines. Now, there was a social as well as a political gulf. The old boys' network of the U.S. Senate was dead — that "fraternal paradise," as *Time* magazine put it, that had "the worst vestiges of a private men's club: unspoken rules, hidden alliances, off-hours socializing and an ethic based at least as much on personal relationships as merit to get things done" — all that was dead and gone.

Yet the women senators, both Republican and Democratic, were determined to develop strong working relationships, even friendships — and, most importantly, to get things done. So, they met regularly for dinner and, when it came to the government shutdown, their relationships led to compromise and the government's reopening for business. As *Time* titled an article on the subject, "Women Are the Only Adults Left in Washington."

It began when Susan Collins, a Republican senator from Maine, put together a three-point plan she thought both parties could agree on. A few days later, GOP senators Lisa Murkowski of Alaska and Kelly Ayotte of New Hampshire signed on. Murkowski had overcome a Tea Party challenge to win reelection, whereas Ayotte benefited from Tea Party support. "I think what our group did was pave the way, and I'm really happy about that," said Senator Collins.

Soon, the GOP trio began working with Barbara Mikulski and Patty Murray on the Democratic side and put together a deal to avert a disastrous default. "In a Senate still dominated by men," the *New York Times* reported, "women on both sides of the partisan divide proved to be the driving forces that shaped a negotiated settlement. The three Republican women put aside threats from the right to advance the interests of their shutdown-weary states and asserted their own political independence." Six of the thirteen senators on the committee trying to work out the debt deal were women, and when the bipartisan deal was finally announced, Sen. John McCain (R-AZ) conceded, "Leadership, I must fully admit, was provided primarily from women in the Senate."

But that was just the beginning. Now that the Senate had put together a budget, it needed to reach an agreement with the House. This meant that Patty Murray, as chairwoman of the Senate Committee on the Budget, had to come to terms with Paul Ryan, chairman of the House Budget Committee — no easy task in the context of a relentlessly partisan, crisis-driven Congress that had not agreed on a budget in years.

Like many compromises, this was a deal in which neither side got everything it wanted. Republicans didn't get the cuts to Medicare and Social Security they sought, and Democrats didn't get the tax hikes on the rich they wanted. Instead of trying for "a grand bargain," Patty and Ryan focused on common ground, and the result was an extraordinarily rare example of bipartisanship during a bitterly dysfunctional period that led to Congress's passing a budget resolution for the first time since 2009. "If we didn't get a deal," Patty said, "we would have faced another continuing resolution that would have locked in the automatic cuts — or worse, a potential government shutdown in just a few short weeks . . . It's a compromise — and that means neither side got everything they wanted, and both sides had to give a bit." As to why women were able to do what men couldn't, Barbara Mikulski put it succinctly: "We did it because we listened to each other and functioned with maximum respect."

Yet another example of women breaking through congressional gridlock took place in 2014 when Debbie Stabenow, then chairwoman of the Senate Agriculture Committee, managed to steer an innovative farm bill through Congress that provided crop insurance to fruit and vegetable farmers and allowed food-stamp recipients to spend more on healthy foods. In short, it was far more amenable to the organic, farm-to-table, and fruit-and-vegetable ethos than the agribusiness-friendly bills of yore had been.

Debbie accomplished this by using an approach similar to Patty Murray and Barbara Mikulski's. "We look for ways to try to find common ground or win-win situations instead of trying to put somebody in their corner and creating a win-lose," Debbie told me. "And I think that's a difference in style between women and men.

"It's not that we're not strong or we're unwilling to fight. I made sure that there was not a nickel cut out of the regular food-assistance programs — that was the bottom line for me. But I also knew I was negotiating with people who had different priorities. So, I approached it from the standpoint of saying, 'Look, this is important for me. That's important to you. Let's cut a deal that does both.' I was looking for ways for people to feel they got a win. That's how you put together a coalition. I think you see women on both sides of the aisle being more practical in these kinds of negotiations."

The farm-bill negotiations also showed that women often have a broader view of what a bill can really mean. Who would have thought a farm bill was really about women and families? But Debbie realized that, in addition to food stamps, which ensured that poor families were fed, it was important to have policies that encouraged good nutrition and health. And to make sure those policies were in the bill, Debbie spent two and a half years wooing Republicans with entirely different interests. She visited Sen. Pat Roberts, who was on the Senate Agriculture Committee, in his home state of Kansas, and then invited Roberts to her state of Michigan. Similarly, Debbie went to Mississippi to woo Sen. Thad Cochran, also on the Agriculture Committee. "It takes time to build those relationships," said Debbie. "It's not my way or the highway, or who's got the biggest muscles. It's about trying to find a way to build trust in a relationship."

Which is precisely what Debbie did. As a result, in February 2014, President Obama signed the farm bill that Debbie spent two and a half years shepherding through Congress, heralding a new era in our national agricultural policy, which now reflected the changing eating habits of American consumers. Organic farmers, fruit growers, and producers of other healthy foods finally began to share in taxpayer dollars that had once gone entirely to gigantic agribusiness corporations. Subsidies to traditional commodities were cut by more than 30 percent over the next ten years, whereas funding for fruits, vegetables, and organic produce increased by more than 50 percent.

MEANWHILE, EVEN THOUGH the 2014 midterms had yet to take place, everyone was already speculating whether Hillary would run in 2016. No one in either party, man or woman, had better credentials. Gallup polls showed her to be the most admired woman in the world year after year. As in 2008, Hillary was already the prohibitive favorite to be the Democratic nominee — and this time there was no Barack Obama in the wings to challenge her. In polls matching her with the vast array of potential GOP nominees, Hillary was the clear winner again and again.

Even though Hillary had yet to declare, the Republicans began to attack — more than two full years before the next presidential elections. As a former secretary of state, Hillary had far better foreign-policy credentials than any Republican candidate on the horizon, so, following the tried-and-true GOP strategy, the Republicans went after those strengths.

More specifically, they criticized Hillary with regard to the 2012 attack by Islamic militants on the U.S. diplomatic compound in Benghazi, Libya, which killed U.S. ambassador Christopher Stevens and three other Americans. For months, the Republicans accused

Hillary of failing to launch a rescue attempt that might have saved lives and of creating a politically motivated campaign after the attack to deceive the American people about what really happened.

By early 2014, however, a bipartisan report by the Senate Intelligence Committee concluded that there was no merit to such allegations. "The Committee has reviewed the allegations that U.S. personnel . . . prevented the mounting of any military relief effort during the attacks, but the Committee has not found any of these allegations to be substantiated," the report said. Nevertheless, Sen. Rand Paul (R-KY) charged that Hillary's failure to send reinforcements to Benghazi "should limit Hillary Clinton from ever holding high office."

(Like the Senate, on July 31, 2014, the House Intelligence Committee also debunked all the allegations against Hillary, concluding, among other things, that "there was no intelligence failure" surrounding the Benghazi attacks, that there was no "stand down order" given to American personnel attempting to offer assistance, and that intelligence assessments were not politically motivated in any way.)

In a more rational world, that might have been the end of it — but it wasn't. By May 2014, after there had already been thirteen public hearings on Benghazi, no fewer than 206 Republicans out of 234 in the House vied for the seven Republican seats on the newly formed House Benghazi Select Subcommittee. It was as if marching orders had gone out to all the Republicans in Congress: we're going after Hillary on Benghazi, and if you want to score points, this is where to do it. On Fox News, it was all Benghazi all the time.

AT THE SAME TIME, the 2014 midterm election cycle was under way. In November, we experienced another period of creeping rather than leaping. In fact, "creeping" is a more-than-generous characterization of the Democrats' progress: in 2014, they lost thirteen seats in the House and nine in the Senate, leaving both houses of Congress in Republican hands.

On the bright side, EMILY's List helped elect nine new women, five of whom were women of color. We also helped Gina Raimondo become governor of Rhode Island and reelected Jeanne Shaheen to the Senate from New Hampshire over Scott Brown, who had moved there from neighboring Massachusetts after his 2012 loss to Elizabeth Warren. On the flip side, two Democratic women lost their Senate seats: Kay Hagan of North Carolina and Mary Landrieu of Louisiana. Kay's loss was terribly disappointing, as she had been a terrific senator for North Carolina. As for Mary, we had disagreed with her support of the Santorum legislation on late-term abortion in 1999, but I had known her so long and I was sorry to see her go.

Another of EMILY's List's major undertakings was our Madam President project for 2016. Just in case Hillary didn't run, we wanted to make sure that there was a woman on the ticket, and by this time, there were plenty of Democratic women with extraordinarily impressive credentials — senators such as Elizabeth Warren, Kirsten Gillibrand, and Amy Klobuchar; governors such as Maggie Hassan (D-NH) and Gina Raimondo (D-RI); and former officials such as Janet Napolitano, who had been secretary of homeland security, attorney general, and governor of Arizona, and who is currently president of the University of California, and Kathleen Sebelius, the former governor of Kansas, who has served as secretary of health and human services from 2009 to 2014.

For male Democrats, it has long been the case that if you were the governor of a red state and a cabinet secretary, you generally made the Democratic short list by default. Our bench was strong enough that we felt women should be getting the same treatment. And our campaign succeeded in that regard, with major national figures Elizabeth Warren, Amy Klobuchar, and Kirsten Gillibrand at the head of the pack.

IN APRIL 2015, JUST a month after EMILY's List's thirtieth-anniversary gala, Hillary announced her candidacy at a stirring event on New York's Roosevelt Island. Her speech was not about running to be the first woman president, but it could have been given only by a woman. That's because it was about advocating for American families — listening to them, putting their needs first, and with absolute determination getting things done for them and their futures. That's what women do.

Throughout the summer, Hillary remained a prohibitive favorite to win the Democratic nomination, and polls still showed her with comfortable leads over various leading contenders for the Republican nomination. At last, we were finally in a position to break through the highest last glass ceiling of all!

But I've been around politics too long to take anything for granted. In June 2015, the House Select Committee on Benghazi — the tenth congressional committee to investigate the events surrounding the attacks — subpoenaed Sidney Blumenthal, a former journalist, Clinton White House adviser, and friend of Hillary Clinton's. At this point, Republicans in Congress had spent more time investigating the assault in Libya than previous governments had inquired into Pearl Harbor, the Kennedy assassination, or the Iran-Contra scandal.

In more than nine hours of closed-door testimony, the Republican-led committee asked Blumenthal only 4 questions about security in Benghazi but more than 160 questions about his relationship and communications with the Clintons. As Adam Smith (D-WA), a member of the Benghazi committee, put it, "It's pretty clear at this point that this is a political investigation focused on Hillary." If the past is any guide, it was equally clear that this was not going to be the end of what promises to be a very, very tough campaign.

If Hillary wins in November 2016, it will be, of course, a huge step for women. But we must remember that if we achieve this once-unimaginable goal — her presidency — it is still just one step on a much bigger journey. Far from marking the end of our accomplishments, it should mark a new beginning. If anything, it should be a call to arms, a powerful reminder of how much we can accomplish and how much more work we have to do. Just ask the women of Israel, as I did when I went there in 1993, after the Year of the Woman. Those women were told again and again, "Hey, you had Golda Meir," who was prime minister from 1969 to 1974. As if that were the answer to everything.

It's not, of course.

To be sure, whether it's Hillary Clinton today or Gloria Steinem and Betty Friedan in the sixties and seventies, it is important to have leaders who will go first, articulate a vision, and chart a path for millions of people to follow. It took a leader like Rosa Parks, "the First Lady of Civil Rights," to refuse to give up her seat in the front section of the bus in Montgomery, Alabama, in 1955. But it also took hundreds of people to march across the Edmund Pettus Bridge in Selma, Alabama, for voting rights. It takes hundreds of lawyers to fight for these rights, and members of Congress to change the laws, and judges to enforce them, and people to send money to activist organizations, and hundreds of thousands more people to act. It takes millions of members of EMILY's List and tens of millions of voters. Without all this, any movement is doomed.

And, it is essential to realize, as I learned over and over again, that in each of these movements social change was never permanent. Progress ebbs and flows. And in doing so, it creates both challenges and opportunities for those who want to see women progress. The ebb and flow also serves as a constant reminder warning us that we should never take the progress we have made for granted. In other words, we can make enormous gains, but they can always slip away.

Always.

We see it in the electoral realm more and more often. As much as we've made progress, women still make up less than 20 percent of Congress. Many of the women we elected in the eighties and nineties are reaching retirement age. Some women, like Tammy Baldwin and Mazie Hirono, leave safe congressional seats to run for higher office. Not all of them win. So, EMILY's List constantly struggles to keep adding new women, sometimes just to protect the gains we've already made.

Which is precisely why it is so important that EMILY's List continue to be a vehicle to give men and women a way to act on their values, for people who think our government should have real diversity, who think that it's wrong that only 19.4 percent of the people in Congress are women, who think our policies would be richer and more effective if Congress represented the female half of our populace. It's why it is so important to build a community of people who believe in social change and progressive values, who fight to elect progressive women to city councils and state legislatures, and who support women to take back the Senate and the House.

Yet women will never have parity until the Republicans open their doors to women. Though women make up more than a third of the Democrats in the House, Republican women represent less than 10 percent of the Republicans in that chamber. That's right: for every Republican woman in the House, there are nine Republican men. In this day and age, that's a national embarrassment, and it impacts the policies Republicans come up with.

Those numbers go a long way toward explaining why the Republicans have such extreme right-wing, antiwoman policies. If they had more women in their caucus, it's hard to imagine that they would be so ham-fisted as to hold a panel on family planning and not allow women to testify. It's hard to imagine they would feel so comfortable espousing their ignorant claims about women's sexuality and issues like gender discrimination in pay and health care. Surely Republicans could learn a great deal if they allowed more women's voices in their deliberations. And that would be good for the country.

I'm certain of this because I can look back on the past thirty years with a profound sense of accomplishment and see exactly how far women have come in politics and how dramatically effective EMILY's List has been. As this chart illustrates, the progress of both parties from 1969 right up until the mid-eighties, when the Republicans took a slight lead over the Democrats in electing women, was minimal.

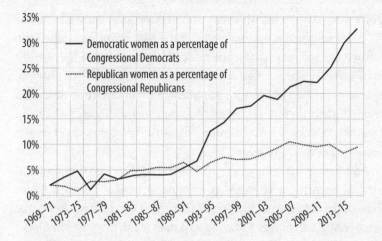

35%
30%
25%
20%
15%
10%
5%
0%

—— Democratic women as a percentage of
    Congressional Democrats
········ Republican women as a percentage of
    Congressional Republicans

1969–71 1973–75 1977–79 1981–83 1985–87 1989–91 1993–95 1997–99 2001–03 2005–07 2009–11 2013–15

But then, in 1985, came EMILY's List — and with it, the beginning of that wonderfully steep hockey-stick curve of the type that venture capitalists love so much. Today, 14 out of 44 Democrats in the United States Senate are women. That's over 30 percent — up from zero when EMILY's List began. Similarly, there are now 62 Democratic women in the House — up from 12 when we began. In all, during those thirty years, we helped elect 110 Democratic women to the House and 19 to the Senate. We elected the first openly gay U.S. senator and the first woman to represent Wisconsin in the Senate, Tammy Baldwin; the first woman senator from Massachusetts, Elizabeth Warren; and the first Asian American woman to serve in the Senate, Hawaii's Mazie Hirono. More than a third of the women we helped elect to Congress are women of color. In New Hampshire, EMILY's List candidates swept the state, giving it the first-ever all-woman congressional delegation. In all, we have changed the percentage of women in the Democratic caucus from the low single digits to 32 percent.

Virtually all the progress of electing women to top political offices has come on the Democratic side. The Republican progress has flatlined for thirty years. It's a scandal. Perhaps even more important, once these women were elected, they made an astounding difference. During our thirtieth-anniversary gala, Amy Klobuchar (D-MN) summed it up succinctly. In the past seven years, she said, the average woman senator introduced ninety-six bills, the average man seventy. The average woman senator had more than nine co-sponsors for her legislation; the average man had fewer than six. Even though there were twenty women in the Senate, when the Democrats were in control nine out of twenty committees in the Senate were chaired by women.

And look at what they have accomplished! I've said that no senator has been more effective in promoting women's health issues than Barbara Mikulski, the first EMILY's List candidate to break the senatorial glass ceiling. Amazing as it may seem, prior to Mikulski's election, the National Institutes of Health had all but ignored women's health in their research. The health needs of the female half of the population were not considered important enough to be worthy of research — because Congress was almost entirely male. But, thanks largely to Senator Mikulski's efforts, in 1990 the National Institutes of Health established the Office of Research on Women's Health to strengthen and enhance research related to diseases such as breast cancer, cervical cancer, and other conditions that affect women. As a result, millions of lives have been saved.

As I've noted, Agriculture Committee chair Debbie Stabenow made sure that the farm bill protected food stamps for families living in poverty. Sen. Patty Murray chaired the budget effort and protected critical programs like Head Start. And, in the spring of 2015, as I watched Senators Claire McCaskill and Kirsten Gillibrand grill four-star generals about what constitutes a serious sexual offense and saw the support coming from the five additional women on the Senate Armed Services Committee, I could not help but think back to the days in 1991 when the Senate Judiciary Committee, with no women at all, shamefully swept the issue of Clarence Thomas's alleged sexual harassment of Anita Hill under the rug, with all of America watching.

Fortunately, the McCaskill-Gillibrand hearing was not a replay of the Thomas-Hill debacle. For decades, thousands of women who were sexually assaulted in the military were belittled and mistreated. For decades, their claims were dismissed out of hand, and the perpetrators got away with rape. For decades, the military pledged reforms. Thanks to the leadership of the women, the military leaders were now given no choice: they had to institute new policies to protect victims and to prosecute perpetrators.

Likewise, on the House side, in 2010 then Speaker Nancy Pelosi and the other Democratic women in the House made certain that the Affordable Care Act — aka Obamacare — not only ended gender discrimination in health insurance but also provided coverage of preventive medical procedures such as mammograms and cervical cancer screenings, and that it included contraception. Rep. Carolyn Maloney (D-NY), with the aid of her sisters in Congress, was finally able to strengthen laws against domestic violence after a long and divisive fight. Rosa DeLauro continues to lead House Democrats on workplace equality issues and on creating policies that help working families, like paid sick leave and family leave.

The obvious conclusion is that creating equal representation for women is as much an imperative today as it was when EMILY's List began. Yet today, many progressives are willing to step away from our mission, as if the job were finished. Every time primary season rolls around, a handful of progressives will examine the field of candidates and ask me, "Well, isn't this guy a better candidate than this woman?"

My answer inevitably is "Probably not." Close examination of the records actually shows that most Democratic-primary candidates agree on virtually every issue. There may be nuanced differences on positions or different issues a candidate has led on, but essentially their similarity is what makes it so difficult for Democratic voters to decide whom to vote for. But electing women candidates brings us one step closer to reaching true diversity in government. If you believe in diversity and equal representation, my advice is simple: all things being equal, vote for the women.

Ultimately, EMILY's List's mission is about much more than implementing a noble but abstract ideal. At a time when tens of millions of women have been entering the workforce, when incomes are stagnating, when families need two paychecks more than ever, it's essential that the rules and regulations of the workplace reflect these changing conditions. As of 2013, in four out of ten households, mothers are the primary breadwinners—either because they are single parents or because they earn as much as or more than their spouses. For tens of millions of children, economic survival depends on the equal and fair treatment of women in the workplace.

And if we look at Congress—bitterly polarized, deliberately gridlocked, willing to shut down the government, determined to obstruct Obamacare, and doing all manner of things that sink its approval rating to as low as 9 percent—one of the reasons it doesn't work well can be explained in four simple words: there aren't enough women. At the end of the day, if half the representatives in Congress were women, it is hard to believe that public education would be disintegrating before our eyes, that women would still be making less money than men, and that the sacred halls of Congress would resemble an acrimonious, highly partisan shouting match that has earned the contempt of the American people.

Women make it work.

We have proved again and again that when women have power, they can make things happen, and when our passions are harnessed, we can change the world. And then we all win.

ACKNOWLEDGMENTS

WHEN WOMEN WIN IS the result of a wonderful collaboration with a terrific writer, Craig Unger. As with all collaborations, we brought to the book different experiences and strengths and melded them into the resulting manuscript. As a result, there are many people we want to thank, both collectively and individually.

In the mid-1980s, a group of passionate feminists joined together with the goal of electing women to the highest levels of political office. Referring to ourselves, somewhat facetiously, as the Founding Mothers, we invested our time and money into building EMILY's List. For their personal friendship, hard work, and political support I am particularly indebted to these organizing members of EMILY's List: Marie Bass, Donna Brazile, Ranny Cooper, Betsy Crone, Irene Crowe, Kathleen Currie, Gail Harmon, Nikki Heidepriem, Joanne Howes, Ann Kolker, Debbie Landau, Judy Lichtman, Mimi Mager, Joan McLean, Jane McMichael, Cat Scheibner, Lael Stegall, and Mary Ann Stein. Among them, Craig and I are especially grateful to Judy, Joanne, Marie, Betsy, Mimi, and Ranny for sharing their memories of the early days of EMILY's List.

EMILY's List was built by the entrepreneurial spirit and dedication of a talented, mission-driven staff. The staff has always inspired me, and I am grateful for the many professional friendships I have built with the men and women who have worked for EMILY's List. I have forged a partnership with a small group of executive directors, who used their political and management savvy to turn EMILY's List into a powerful full-service political organization. I am particularly grateful for the friendship and commitment of these executive directors and for the time they have spent being interviewed for this book: Rosa DeLauro, Wendy Sherman, Mary Beth Cahill, Joe Solmonese, and Ellen Moran.

CRAIG AND I ALSO thank the following staff members for being interviewed about their time at EMILY's List: Denise Feriozzi, Sheila O'Connell, Martha McKenna, Jonathan Parker, Jen Pihlaja, Karin Johanson, Kate Coyne-McCoy, Chris Esposito, Heather Colburn, Simone Ward, Amy Green, Mitchell Lester, and Allison Dale-Riddle. I especially thank Jeanne Duncan for many years of collaboration, for being "EMILY's archivist," and, along with Kate Black, for helping us with fact checking.

I especially thank the women who helped create the major-donor program of EMILY's List. For many years, we traveled together and shared mishaps, hopes, frustrations, and joys. I will always be grateful for sharing this adventure with my friends on "the Team": Judi Kanter, Judy Loeb Goldfein, Anna Lidman, Shellie Sachs Levin, Sherry Merfish, and Patricia Williams.

In February 2010, I turned over the presidency of EMILY's List to Stephanie Schriock. It has been wonderful for me to watch as EMILY's List has grown and prospered under her leadership. Stephanie and I have forged a strong relationship over these five years, and I'm appreciative of her friendship and belief in the mission of EMILY's List. Craig and I thank Stephanie and some of the current staff of EMILY's List who have helped us, particularly Jess O'Connell, Kate Black, Jess McIntosh, Marcy Stech, Liana Eisman, and Kate Watt.

Craig and I are both are extremely grateful to the elected women we interviewed for their candor and encouragement, including Sen. Tammy Baldwin, Secty. Hillary Clinton, Rep. Rosa DeLauro, Sen. Barbara Mikulski, Rep. Gwen Moore, Gov. Janet Napolitano, Leader Nancy Pelosi, Gov. Kathleen Sebelius, and Sen. Elizabeth Warren.

In addition, this book would not have been possible without the help of many people at Houghton Mifflin Harcourt. We have had the privilege of being edited by Deanne Urmy, and throughout, her editorial judgment has been superb. We are also grateful to Michelle Triant and Ayesha Mirza at HMH, who handled the publicity and marketing aspects of the book with the highest level of professionalism. Our thanks also go out to Jenny Xu for manuscript help, Tammy Zambo for careful copyediting, and Jessica Resnick for proofreading.

Craig also thanks his agents on this book, Madeleine Morel of 2M Communications and Sloan Harris of International Creative Management, who were enormously helpful. And he appreciated enormously the friendship of Sidney and Jackie Blumenthal, whom he thanks for their generosity and hospitality during his visits to Washington. Thanks also to Howard Dickler for kindly lending Craig his upstate New York country retreat.

Craig and I thank Donald Sussman for his generous grant supporting this book, and David Thau at the American Independent Institute for facilitating it.

I thank my agents on this book, David Kuhn and Lauren Sharp of Kuhn Projects, for guiding me through the mysterious world of publishing. I thank Phillip McLaughlin and Frank Bille for their business acumen and friendship, which made it possible for me to devote time to this book. I thank Steve Rosenthal, Harold Ickes, Gina Glantz, Andy Stern, and Cecile Richards for their partnership and friendship, especially during the hectic adventure that we shared in the 2004 election. And I thank Debbie Walsh and the Center for American Women and Politics for being an important resource for me for thirty years.

I have always felt a heartfelt connection to the members and staff of EMILY's List and the candidates we support. Together, we share a deep belief in our democracy and in the important contribution of women in the leadership of our country. We have come together through EMILY's List to join our passion and political vision with generations of progressive activists and leaders who have shaped our country. It is my honor to have been partners with so many generous men and women who have the highest ideals and who follow through on their beliefs with investments of time and money. You have made the world a different place. You have shown how our country benefits when women win. My deepest gratitude goes to all of you.

Ellen R. Malcolm

NOTES

This book is based primarily on Ellen Malcolm's experiences and observations, as well as newsletters in EMILY's List's archives, which were ably edited and compiled by Jeanne Duncan.

In addition, much of the information comes from firsthand interviews conducted by Craig Unger and Ellen with many of Ellen's friends and colleagues in the political world, including Tammy Baldwin, Marie Bass, Mary Beth Cahill, Hillary Clinton, Betsy Crone, Rosa DeLauro, Judy Goldfein, Jane Hickie, Joanne Howes, Karin Johanson, Judi Kanter, Judith Lichtman, Sherry Merfish, Barbara Mikulski, Gwen Moore, Nancy Pelosi, Jennifer Pihlaja, Stephanie Schriock, Wendy Sherman, Joe Solmonese, Debbie Stabenow, Jennifer Treat, and Elizabeth Warren.

In addition, while on assignment for the *New Republic* in 1990, Craig Unger interviewed Ann Richards and Molly Ivins, among others, for an article on Richards's successful gubernatorial race. The notes that follow refer only to information we researched in books, journals, newspapers, and the Internet.

INTRODUCTION: The Last Glass Ceiling

page

ix *In 1985, 5 percent:* "History of Women in the U.S. Congress," Center for American Women and Politics, http://www.cawp.rutgers.edu/history-women -us-congress.

There were just two women: Ibid.

x *our representative democracy:* In 1985, there were 115,730,000 men and

122,194,000 women in the United States, making women 51.358 percent of the total population. National Center for Health Statistics, Centers for Disease Control and Prevention, "Population by Age Groups, Race, and Sex for 1960–97," http://www.cdc.gov/nchs/data/statab/pop6097.pdf.

xi *the first woman of color to be elected:* "Ayanna Pressley," City of Boston, http://www.cityofboston.gov/citycouncil/councillors/pressley.asp.

xiv *"put on your lipstick":* Annie-Rose Strasser, "After GOP Blocks Pay Equity, Sen. Barbara Mikulski Calls on Women to Start a 'New American Revolution,'" *ThinkProgress,* June 5, 2012, http://thinkprogress.org/politics/2012/06/05/495447/mikulski-pay-equity-revolution.

But the day before: Ibid.

the longest-serving woman: Justin Wortland, "Barbara Mikulski, Longest-Serving Woman in Congress, to Retire," *Time,* March 2, 2015, http://time.com/3728249/barbara-mikulski-retires/.

"Do I spend my time raising money": Brian Naylor, "Sen. Mikulski, Groundbreaker for Female Legislators, Won't Seek Re-election," NPR.org, March 2, 2015, http://www.npr.org/2015/03/02/390245024/sen-mikulski-ground-breaker-for-women-legislators-wont-seek-reelection.

"I'm gonna be around": Seung Min Kim, "How Barbara Mikulski Led the Way for Women in Congress," *Politico,* March 2, 2015, http://www.politico.com/story/2015/03/barbara-mikulski-congress-retirement-115681.html.

xv *We helped Carol Moseley Braun:* "History of Women in the U.S. Congress," Center for American Women and Politics, http://www.cawp.rutgers.edu/history-women-us-congress.

We helped Jeanne Shaheen: "About Jeanne," Jeanne Shaheen biography, http://www.shaheen.senate.gov/about/biography/.

1. A Political Education

1 *McCarthy won 71 percent:* "Our Campaigns — PA US President — D Primary Race — Apr. 23, 1968," Our Campaigns.com. http://www.ourcampaigns.com/RaceDetail.html?RaceID=35989.

2 *In the aftermath of the King assassination:* Clay Risen, "The Legacy of the 1968 Riots," *The Guardian,* April 4, 2008, http://www.theguardian.com/commentisfree/2008/apr/04/thelegacyofthe1968riots.

3 *In August, the McCarthy campaign:* James C. Kitch, "Skip Convention, McCarthy Pleads," *Harvard Crimson,* August 13, 1968, http://www.thecrimson

.com/article/1968/8/13/skip-convention-mccarthy-pleads-psenator
-eugene/.

4 *In 1971, Common Cause sued:* Lawrence M. Salinger, *Encyclopedia of White-
Collar and Corporate Crime* 2nd ed. (Thousand Oaks, Calif.: Sage, 2005), 1:122.

5 *Shortly after Nixon's departure:* "The FEC and the Federal Campaign Finance
Law," Federal Election Commission, February 1, 2014, http://www.fec.gov/
pages/brochures/fecfeca.shtml.

Congress cut funding to South Vietnam: Lauren Zanolli, "What Happened
When Democrats in Congress Cut Off Funding for the Vietnam War?,"
History News Network, April 13, 2007, http://historynewsnetwork.org/
article/31400.

6 *In 1963,* The Feminine Mystique: Betty Friedan, *The Feminine Mystique,* 20th-
ann. ed. (New York: Norton, 1983).

In 1968, the word sexism: *Merriam-Webster Online,* s.v. "sexism," http://www.
merriam-webster.com/dictionary/sexism.

In 1970, Kate Millett's Sexual Politics: Kate Millett, *Sexual Politics* (Garden
City, N.Y.: Doubleday, 1970); Germaine Greer, *The Female Eunuch* (London:
MacGibbon & Kee, 1970).

"could boil the fat off a taxicab driver's neck": Laura Mansnerus, "Bella Abzug,
77, Congresswoman and a Founding Feminist, Is Dead," *New York Times,*
April 1, 1998, http://www.nytimes.com/learning/general/onthisday/bday/
0724.html.

"this woman's place is in the House," Susan Baer, "Founding, Enduring Feminist
Bella Abzug Is Dead at 77: 'Battling Bella' Served Three Terms in House,"
Baltimore Sun, April 1, 1998, http://articles.baltimoresun.com/1998-04-01/
news/1998091029_1_bella-abzug-feminist-woman-in-congress.

7 *the "click" was the "moment of truth":* Jane O'Reilly, "The Housewife's Moment
of Truth," *New York Magazine,* December 20, 1971, http://nymag.com/news/
features/46167/index4.html.

2. A Movement Begins

11 *That last groundbreaking decision:* Steven Greenhouse, "Women of
Distinction," *New York Times,* April 5, 2004, http://www.nytimes.
com/2004/04/05/us/mildred-jeffrey-93-activist-for-women-labor-and
-liberties.html; Harold Meyerson, "Unsung Heroine," *American Prospect,* April
2, 2004, http://prospect.org/article/unsung-heroine.

12 *Smith suggested that the word* sex: Jamie Malanowski, *Washington Post,* February 6, 1994, http://www.washingtonpost.com/archive/opinions/1994/02/06/racists-for-feminism-the-odd-history-of-the-civil-rights-bill/e9363c95-d93b-4b40-a0f6-e1343e1b1c59/.

In fact, when Smith introduced the wording change: Clay Risen, "The Accidental Feminist," *Slate,* February 7, 2014, http://www.slate.com/articles/news_and_politics/jurisprudence/2014/02/the_50th_anniversary_of_title_vii_of_the_civil_rights_act_and_the_southern.html.

they feared such laws would jeopardize: Louis Menand, "The Sex Amendment: How Women Got In on the Civil Rights Act," *The New Yorker,* July 21, 2014, http://www.newyorker.com/magazine/2014/07/21/sex-amendment.

13 *Conversely, the biggest supporters of the ERA:* Ibid.

"implacable determination, a lawyer's grasp": Harold Jackson, "Martha Griffiths," *The Guardian,* April 28, 2003, http://www.theguardian.com/news/2003/apr/29/guardianobituaries.haroldjackson.

Social Security, she discovered: Elaine Woo, "Martha Griffiths, 91; Pioneering Politician Pushed ERA, Sex Bias Ban Through Congress," *Los Angeles Times,* April 25, 2003, http://articles.latimes.com/2003/apr/25/local/me-griffiths25.

"What are you running": Jackson, "Martha Griffiths."

14 *He knew all about equal rights:* Menand, "Sex Amendment."

15 *When the National Women's Political Caucus:* Donna Radcliffe and Jacqueline Trescott, "Carter and the Women's Caucus," *Washington Post,* March 31, 1977.

17 *Margaret Heckler, the longtime "Rockefeller Republican":* Donnie Radcliffe, "The Women's Caucus," *Washington Post,* April 27, 1978, http://www.washingtonpost.com/pb/archive/lifestyle/1978/04/27/the-womens-caucus/be0c45df-ed7f-4b9d-8e97-d3be5cf4e8b1/?resType=accessibility.

Republican senator Nancy Kassebaum (R-KS) championed: Diana Reese, "Kansas Political Daughter Sen. Nancy Landon Kassebaum Baker Rode Father's Coattails," *Washington Post,* July 30, 2013, http://www.washingtonpost.com/blogs/she-the-people/wp/2013/07/30/kansas-political-daughter-sen-nancy-landon-kassebaum-baker-rode-fathers-coattails/.

In 1976, there were none: Roberta W. Francis, "The History Behind the Equal Rights Amendment," National Council of Women's Organizations, http://www.equalrightsamendment.org/history.htm.

"Schlafly herself fit:" Tanya Melich, *The Republican War Against Women: An Insider's Report from Behind the Lines* (New York: Bantam, 2009), Kindle ed.

18 *"Women's libbers are trying"*: Phyllis Schlafly, quoted in ibid.

Schlafly argued — wrongly: Phyllis Schlafly, " 'Equal Rights' for Women: Wrong Then, Wrong Now," *Los Angeles Times,* April 8, 2007, http://www.latimes.com/la-op-schafly8apr08-story.html.

One Schlafly ally cautioned: Melich, *Republican War Against Women.*

One Republican woman said: Ibid.

In July 1978, NOW organized: Francis, "History Behind the Equal Rights Amendment."

19 *Rep. Barbara Jordan (D-TX) became:* Barbara Johnson, "Barbara Jordan Keynote Address to 1976 Democratic National Convention," filmed July 12, 1976, YouTube, posted October 19, 2009, https://www.youtube.com/watch?v=sKfFJc37jjQ.

Rep. Barbara Mikulski (D-MD) chastised: Susan J. Carroll and Barbara Geiger-Parker, "History of Women in the U.S. Congress," Center for American Women and Politics, Rutgers University, 1983, http://www.cawp.rutgers.edu/history-women-us-congress.

A week after the caucus sent: Janet M. Martin, *The Presidency and Women: Promise, Performance, and Illusion* (College Station: Texas A&M University Press, 2009), 204.

20 *Soon, the number of women:* Ibid., 212.

In the judiciary, Carter appointed: Mary L. Clark, "Carter's Groundbreaking Appointment of Women to the Federal Bench: His Other 'Human Rights' Record," *Journal of Gender, Social Policy, and the Law* 11, no. 3 (2003): 1131–63, http://digitalcommons.wcl.american.edu/cgi/viewcontent.cgi?article=1323&context=jgspl.

3. Starting from Zero

23 *"There are two things":* Michael Beschloss, "Money and Politics Go Hand-in-Hand," ABC News Network Online, May 23, 2015, http://abcnews.go.com/Politics/story?id=121651.

25 *Millie Jeffrey was defeated:* Andy Lippman, "Domestic News," Associated Press, July 15, 1979.

26 *when girls as young as ten:* Thomas Dublin, "Women, Work, and Protest in the Early Lowell Mills: 'The Oppressing Hand of Avarice Would Enslave Us,' " *Labor History* 16, no. 1 (1975): 99–116.

27 *Esther opposed the Equal Rights Amendment:* Judy Klemesrud, "Assessing Eleanor Roosevelt as a Feminist," *New York Times,* November 5, 1984, http://www.nytimes.com/1984/11/05/style/assessing-eleanor-roosevelt-as-a -feminist.html.

28 *The subject of a stinging rebuke:* "GOP Leader, Feminist Mary Crisp Dies at 83," *Washington Post,* April 26, 2007, http://www.washingtonpost.com/wp-dyn/ content/article/2007/04/25/AR2007042502882.html.

"We will work for the appointment": Lauren Feeney, "Timeline: The Religious Right and the Republican Platform," Moyers & Company, August 31, 2012, http://billmoyers.com/content/timeline-the-religious-right-and-the -republican.

29 *Tim LaHaye and his wife, Beverly:* "Our History," Concerned Women for America, http://www.cwfa.org/about/our-history/.

32 *As a relative of Sen. Howard Metzenbaum:* Gary A. Tobin, with Sharon L. Sassler, *Jewish Perceptions of Antisemitism* (New York: Plenum Press, 1988), 62.

Harriett was the only female Democrat: "Summary of Women Candidates for Selected Offices, 1970–2014 (Major Party Nominees)," Center for American Women and Politics, Rutgers University, 2014, http://www.cawp.rutgers.edu/ sites/default/files/resources/can_histsum.pdf.

there had been a grand total of eight: Doris Weatherford, *Women in American Politics: History and Milestones* (Thousand Oaks, Calif.: CQ Press, 2012), 197.

"If it takes lynching": "Felton, Rebecca Latimer," History, Art, and Archives: United States House of Representatives, http://history.house.gov/People/ Detail/13054.

Similarly, in 1936, Rose Long: "Long, Rose McConnell," History, Art, and Archives: United States House of Representatives, http://history.house.gov/ People/Detail/17125.

The following year, Alabama governor Bibb Graves: "Graves, Dixie Bibb," History, Art, and Archives: United States House of Representatives, http:// history.house.gov/People/Detail/14057.

4. The Vicious Cycle (and How to Break It)

36 *In late September:* Adam Clymer, "Danforth Appears to Lead in Missouri Senate Contest," *New York Times,* October 13, 1982.

She won backing: Evans Witt, "Woods, and Women's Groups, in Uphill Fight Against Danforth," Associated Press, October 14, 1982.

37 *Thanks to her experience in TV news:* Elisabeth Bumiller, "Senate Showdown in the Show-Me State; Harriett Woods and John Danforth: No Longer Polls Apart," *Washington Post,* October 15, 1982.

the Saint Louis Globe-Democrat*'s poll:* Ibid.

Now that he finally felt the heat: Nathaniel Sheppard, "Poll Finds Danforth and His Opponent in Missouri Running Neck and Neck," *New York Times,* October 25, 1982.

"I can't prove it, of course,": David S. Broder, "Daily Polls Helped GOP Keep Senate Edge," *Washington Post,* November 7, 1982.

5. The Ferraro Factor

43 *In large part, that was thanks:* Geraldine Ferraro, with Linda Bird Francke, *Ferraro, My Story* (Evanston, Ill.: Northwestern University Press, 1985), 72.

44 *Pat Schroeder of Colorado:* Ibid., 71.

On September 24: Ibid., 73.

45 *In November, Joan McLean: Geraldine Ferraro: Paving the Way,* directed by Donna Zaccaro, televised by Showtime Networks in March 2014. A DVD was later released by Specialty Studios (San Francisco, Calif.).

Then, as the meal came to an end: Ferraro, *Ferraro, My Story,* 71.

"Am I the only woman": Ibid., 73.

According to Ford's document: "Election Extra," *Newsweek,* November–December 1984.

46 *"When Pat Schroeder, Geraldine Ferraro":* Elisabeth Bumiller, "Four Possible Candidates," *Washington Post,* April 29, 1984.

"If you're looking for a VP, she's the one": Zaccaro, *Geraldine Ferraro.*

"Her name is Geraldine Ferraro": Quoted in Associated Press, "Rep. Ferraro Is Choice of O'Neill for No. 2 Spot," *New York Times,* May 5, 1984.

It was, as Ferraro later put it: R. Cort Kirkwood, "Former Rep. Geraldine Ferraro Dies," *New American,* March 26, 2011, http://www.thenewamerican.com/usnews/politics/item/5684-former-rep-geraldine-ferraro-dies.

Instead, NOW had made: Tom Fiedler, "NOW's Message: 'Woman VP Now,'" *Miami Herald,* June 30, 1984.

47 *"Today, I'm delighted to announce":* Bernard Weinraub, "Geraldine Ferarro Is Chosen by Mondale as Running Mate, First Woman on Major Ticket," *New York Times,* July 12, 1954, http://www.nytimes.com/learning/general/ onthisday/big/0712.html.

"Let me say that again": "Excerpts from Ferraro Speech on Foreign Policy to World Affairs Council," *New York Times,* July 13, 1984, http://www.nytimes .com/1984/07/13/us/excerpts-from-ferraro-speech-on-foreign-policy-to -world-affairs-council.html.

48 *"If we can do this":* Harriet Sigerman, *The Columbia Documentary History of American Women Since 1941* (New York: Columbia University Press, 2003), 385.

"It was one of the most electric moments": Zaccaro, *Geraldine Ferraro.*

"Even my Republican friends": Ibid.

"Throughout the campaign, gender intruded": Maureen Dowd, "Ferraro Campaign: Perspectives That Startle," *New York Times,* October 10, 1984.

49 *voters heard a vice presidential candidate:* Douglas Martin, "She Ended the Men's Club of National Politics," *New York Times,* March 26, 2011, http://www. nytimes.com/2011/03/27/us/politics/27geraldine-ferraro.html.

"We were in North Oaks": Zaccaro, *Geraldine Ferraro.*

She got up and stormed out: Ibid.

51 *In the end, Bush had to fall back:* Gerald M. Boyd, "Aide to Ferraro Demands Bush Make Apology," *New York Times,* October 14, 1984, http://www.nytimes. com/1984/10/14/us/aide-to-ferraro-demands-bush-make-apology.html.

"There has been more emphasis": Maureen Dowd, "Political Outlook Dims for Women After Hopes Raised by Ferraro's Bid," *New York Times,* November 1, 1984.

So, even though seventeen: "Women Candidates for U.S. House Seat: 1976–2014," Center for American Women and Politics, Rutgers University, http://www.cawp.rutgers.edu/fast_facts/elections/documents/canwincong _histsum.pdf.

6. The Founding Mothers

62 *The Catholic daughter:* Lynne E. Ford, *Encyclopedia of Women and American Politics* (New York: Facts on File, 2008), 308.

"Poverty was one thing": Colman McCarthy, "Stick-Shift Populist," *Washington Post*, April 5, 1986.

So, instead of becoming a nun: Marc Fisher and Jenna Johnson, "Mikulski, a Role Model for Generations of Women in Politics, to Retire in 2016," *Washington Post*, March 2, 2015, http://www.washingtonpost.com/local/dc-politics/mikulski-a-role-model-for-generations-of-women-in-politics-to-retire-in-2016/2015/03/02/c6770396-c0ef-11e4-9ec2-b418f57a4a99_story.html.

64   *"Are you kidding?"*: EMILY's List *Notes from Emily Newsletter,* distributed to EMILY's List mailing list, December 1987.

"I didn't want to be": David Maraniss, "The Battle over Unisex Insurance," *Washington Post*, July 18, 1983.

In mid-May 1983: Associated Press, "Mikulski to Offer Party's Response," *Easton (Md.) Star-Democrat*, May 12, 1983, http://stardem.newspapers.com/newspage/91888202/.

After being touted: Associated Press, "Women Candidates Get Credibility Boost," *Boca Raton (Fla.) News*, October 5, 1984, https://news.google.com/newspapers?nid=1291&dat=19841005&id=cBJUAAAAIBAJ&sjid=vIwDAAAAIBAJ&pg=3417,890321&hl=en.

65   *At the time, about two out of three*: Herbert C. Smith and John T. Willis, *Maryland Politics and Government: Democratic Dominance* (Lincoln: University of Nebraska Press, 2012), 55.

7. A Distant Kingdom

71   *In Washington, Anne Bingaman*: Amanda Spake, "Women Can Be Power Brokers Too," *Washington Post Magazine*, June 5, 1988, http://www.washingtonpost.com/archive/lifestyle/magazine/1988/06/05/women-can-be-power-brokers-too/15916940-dbd9-459a-86a7-dfc94b41f469.

72   *The savings-and-loan scandal*: Tom Kenworthy and Donald P. Baker, "Harry Hughes Undergoes Trial by Dilemma; Maryland Savings and Loan Problems Will Figure in Senate Race Decision," *Washington Post*, August 25, 1985.

the "incredible shrinking governor": Michael White, "Senate Awaiting Resolute Radical," *The Guardian*, September 10, 1986.

By mid-September, Barnes: R. H. Melton, "Barnes Hints He's Ready for Senate Run," *Washington Post*, September 17, 1985.

Two weeks later, as had long been expected: David Espo, "Mathias' Retirement Will Hurt GOP," Associated Press, September 29, 1985.

A Baltimore Sun *poll showed her:* Kenworthy and Baker, "Harry Hughes Undergoes Trial by Dilemma."

73 *On Monday, October 28, 1985:* Sandra Sugawara, "Mikulski Opens Senate Race," *Washington Post,* October 29, 1985, http://www.washingtonpost.com/archive/local/1985/10/29/mikulski-opens-senate-race/d7ef1d06-fd46-4092-a059-eeef0279dbfa/.

74 *In November 1985, a WBAL-TV survey:* Sandra Sugawara, "Polls Give Mikulski Big Lead in Democratic Senate Race," *Washington Post,* March 4, 1986, http://www.washingtonpost.com/archive/local/1986/03/04/polls-give-mikulski-big-lead-in-democratic-senate-race/0ccf64e0-6661-4317-9ce3-fcceed9d7776/.

Worse, Barnes had more than twice: Sandra Sugawara and Michel McQueen, "Political War Chests Begin to Grow in Area Campaigns," *Washington Post,* February 1, 1986.

75 *"quick quotable wit, strong stands":* Michel McQueen and Sandra Sugawara, "Feisty Mikulski Surprises Her Doubters Once Again," *Washington Post,* December 15, 1985.

76 *"It's the death by a thousand cuts":* Kenworthy and Baker, "Harry Hughes Undergoes Trial by Dilemma."

nothing he did helped him: Sandra Sugawara, "Barnes Gets Publicity, but Will It Win Him Votes?," *Washington Post,* April 16, 1986.

When the numbers came in: Sandra Sugawara and Michel McQueen, "Mikulski Surges in Md. Fund Raising," *Washington Post,* February 1, 1986.

"front-runner": Ibid.

Less than a week later: Sandra Sugawara, "Several Black Groups Back Mikulski Bid," *Washington Post,* July 10, 1986.

77 *By May, some four months before:* "The Best Candidate: Correction Appended," *New York Times,* May 22, 1986.

both President Reagan: Bernard Weinraub, "The Political Campaign; Reagan Begins 13-State Campaign Tour," *New York Times,* October 24, 1986, http://www.nytimes.com/1986/10/24/us/the-political-campaign-reagan-begins-13-state-campaign-tour.html.

Vice President George H. W. Bush: Lawrence H. Larsen, *A History of Missouri,* vol. 6, *1953 to 2003* (Columbia: University of Missouri Press, 2004), 141.

According to one poll after another: "Poll Says Woods, Bond Very Close," *Cape Girardeau (Mo.) Southeast Missourian,* November 7, 1985, https://news

.google.com/newspapers?nid=1893&dat=19851107&id=G1ogAAAAIBAJ&sjid=TtcEAAAAIBAJ&pg=2829,1013053&hl=en.

78 *According to Don Sipple:* James Wolfe, "Book Details Turning Point in the Campaign," *Cape Girardeau (Mo.) Southeast Missourian,* September 23, 1988, https://news.google.com/newspapers?nid=1893&dat=19880923&id=MrEfAAAAIBAJ&sjid=Q9cEAAAAIBAJ&pg=1553,2716547&hl=en.

The ad, which was put together: Ibid.

According to the book Madam President: Eleanor Clift and Tom Brazaitis, *Madam President: Shattering the Last Glass Ceiling* (New York: Scribner, 2000), 93.

79 *All over the state:* Paul Taylor, "In Missouri Race, It's the Year of the Ad; Empathy Becomes an Issue as Bond and Woods Battle for the Senate," *Washington Post,* August 4, 1986.

In Columbia, Missouri: Ibid.

"She should have had a Marlboro man": James Wolfe, "Book Details Turning Point in the Campaign," *Cape Girardeau (Mo.) Southeast Missourian,* September 23, 1988.

By September, Bond was 8 points ahead: Paul Taylor, "Odds Stay Even in Battle for Senate; Contest for Control Focuses on 16 States but Lacks National Theme," *Washington Post,* September 1, 1986.

80 *"I have decided to make this endorsement":* United Press International, August 8, 1986.

to hope "for a real natural disaster": Michel McQueen and Sandra Sugawara, "Mikulski Keeps Lead in Senate Race; Barnes, Hughes Trailing in Polls," *Washington Post,* September 1, 1986.

A longtime savvy political operative: Martin Weil, "Anne Wexler, Political Adviser and Lobbyist, Dies at 79," *Washington Post,* August 8, 2009, http://www.washingtonpost.com/wp-dyn/content/article/2009/08/08/AR2009080800058.html.

"It depends how tight": Lois Romano, "Women and the Narrow Corridors; At the White House, Frustrations Linger over the Lack of Access and Recognition," *Washington Post,* February 4, 1986.

81 *When all was said and done:* E. J. Dionne, "Maryland Voters Choose 2 Women," *New York Times,* September 10, 1986.

To which Barbara replied: Maureen Dowd, "Razor-Edged Race for Maryland Seat," *The New York Times,* October 21, 1986, http://www.nytimes.com/1986/10/21/us/razor-edged-race-for-maryland-seat.html.

82   *"She spent fifteen years":* Tom Kenworthy, "Mikulski Returns to Her Roots for Rousing Victory Celebration; New Senator, Long Underestimated, Proves Political Mettle," *Washington Post,* November 5, 1986.

82   *But, even before November:* John B. O'Donnell, "Mikulski Expected to Run for Senate Leadership Post," April 22, 1994, http://articles.baltimoresun.com/1994 -04-22/news/1994112109_1_mikulski-senate-leadership-positions.

8. A Star Is Born

85   *Many senators had expected Barbara:* Eric Pianin, "The Abrasive Lady from Baltimore Polishes Her Act," *Washington Post,* June 14, 1987.

"Just like the singer Tom Jones": Ibid.

86   *"I find her style":* Ibid.

87   *But when she decided:* "Women Candidates for U.S. House Seats: 1976–2014," Center for American Women and Politics, Rutgers University, http://www .cawp.rutgers.edu/fast_facts/elections/documents/canwincong_histsum.pdf.

89   *Given that 98 percent:* "Reelection Rates over the Years," Center for Responsive Politics, https://www.opensecrets.org/bigpicture/reelect.php.

As a young girl: Stephanie Salmon, "Ten Things You Didn't Know About Nancy Pelosi," *U.S. News and World Report,* November 7, 2006, http://www .usnews.com/news/articles/2006/11/07/10-things-you-didnt-know-about -nancy-pelosi.

90   *In early 1987, Rep. Sala Burton:* Johnny Miller, "Sala Burton Endorses Pelosi to Replace Her, 1987," *San Francisco Chronicle,* January 22, 2012, http://www .sfgate.com/entertainment/article/Sala-Burton-endorses-Pelosi-to-replace -her-1987-2637521.php.

94   *A few hours after Ann was born:* Craig Unger, "How to Beat Bubba," *New Republic,* October 22, 1990.

In later years, Ann made fun: Donald P. Myers, "The Wit and Grit of Ann Richards," *Newsday,* July 18, 1988.

95   *In 1972, she worked on:* Sarah Weddington, "Ann Richards' Legacy," *Texas Observer,* October 10, 2012, http://www.texasobserver.org/ann-richards-legacy/.

In 1975, David Richards: J. Michael Kennedy and Stuart Silverstein, "Ann Richards, 73; Former Governor of Texas Was Known for Her Freewheeling

Oratory," *Los Angeles Times,* September 14, 2006, http://articles.latimes
.com/2006/sep/14/local/me-richards14.

96 *It was an unusual job:* Michael Holmes, "For Ann Richards, Politics and
Humor Mix with AM-Texas Governor, BJT," Associated Press, April 10, 1990,
http://www.apnewsarchive.com/1990/For-Ann-Richards-Politics-and-Humor
-Mix-With-AM-Texas-Governor-Bjt/id-f002f0c65a6b0c1f3aa2344b7dce3545.

97 *Before long, a "sorry, no-account sumbitch" judge:* Ibid.

   When Ann appeared: Molly Ivins, *Molly Ivins Can't Say That, Can She?* (New
   York: Vintage, 1991), 76.

   "Are there any other white men": J. Michael Kennedy, "1998 Democratic
   National Convention; To Ann Richards, This Evening Was Made for Dreams,"
   Los Angeles Times, July 18, 1988.

98 *In June, the* Washington Post Magazine: Amanda Spake, "Women Can Be
Power Brokers, Too; How Ellen Malcom Learned to Influence Elections — and
Love It," *Washington Post Magazine,* June 5, 1988, http://www.highbeam.com/
doc/1P2-1260942.html.

99 *In the* Washington Post: Lois Romano, "Keynoter Ann Richards, with Her Wit
About Her," *Washington Post,* July 18, 1988.

100 *"You feel like they've made":* Jan Reid, *Let the People In: The Life and Times of
Ann Richards* (Austin: University of Texas Press, 2012), 180, Kindle edition.

   "I wanted to speak": Ibid.

101 *"Walter," she said:* Ibid.

104 *As the* Houston Chronicle *wrote:* Sonia Smith, "Remembering the Last Texan
to Deliver the DNC Keynote," *Texas Monthly,* January 21, 2013, http://www
.texasmonthly.com/story/remembering-last-texan-deliver-dnc-keynote-
address/page/0/1.

9. How to Beat Bubba

108 *And now, at a time when:* Hilary Hylton, "Much More Than a Good Ole
Girl," *Time,* September 14, 2006, http://content.time.com/time/nation/
article/0,8599,1535050,00.html.

109 *The widow of mountaineer Willi Unsoeld:* "Unsoeld, Jolene," History, Art, and
Archives: United States House of Representatives, http://history.house.gov/
People/Detail/20851.

110 *Then, about three weeks before the election:* James Feron, "A Quickening Congressional Race," *New York Times,* October 23, 1988.

With DioGuardi on the defensive: "For Congress from New York," *New York Times,* October 27, 1998, http://www.nytimes.com/1988/10/27/opinion/for -congress-from-new-york.html.

111 *Similarly, Nita eked by:* "DioGuardi Concedes Defeat to Lowey in Re-election Bid," *New York Times,* November 17, 1988, http://www.nytimes.com/1988/11/17/ nyregion/dioguardi-concedes-defeat-to-lowey-in-re-election-bid.html.

Fourteen Democratic women: "Women Candidates for Congress, 1974–2014," Center for American Women and Politics, Rutgers University, http:// www.cawp.rutgers.edu/sites/default/files/resources/canwincong_histsum .pdf.

116 *Her Democratic rivals included:* Celia Morris, *Finding Celia's Place* (College Station: Texas A&M University Press, 2000), 68.

That was largely because of Mattox: Molly Ivins, *Molly Ivins Can't Say That, Can She?* (New York: Vintage, 1991), 76.

All of which left Ann: Maralee Schwartz and David S. Broder, "Readiness to Use Death Penalty Is a Theme of Texas Campaign," *Washington Post,* February 20, 1990.

117 *"Don't go trying to be friendly with me":* David Maraniss, "In Governor's Race, Finger-Pointing Foes," *Washington Post,* July 3, 1989.

"There is sufficient information": Michael Holmes, "Ann Richards Used Drugs, Is Ducking Debates, Opponent Says," Associated Press, March 28, 1990.

118 *"She was afraid"* Morris, *Finding Celia's Place.*

"I said just what I intended": Ibid.

Mattox and White: "Poll Shows Democratic Logjam, Republican Runaway for Governor," Associated Press, March 11, 1990.

119 *Mark White was no longer:* "In Texas Race, a Charge of 'Himmler' Tactics," *New York Times,* April 4, 1990, http://www.nytimes.com/1990/04/04/us/in -texas-race-a-charge-of-himmler-tactics.html.

"so nasty it would gag a buzzard": Dan Rather, quoted in Eleanor Clift and Tom Brazaitis, *Madam President: Shattering the Last Glass Ceiling* (New York: Scribner, 2000), 93.

120 *In July, two Christian activists: Times* Staff, "Political Briefing," *Los Angeles Times,* August 6, 1990.

121 *Allegations that he used to stage:* Roberto Suro, "Texan Hunkers Down After Stumbling on Tongue," *New York Times,* May 19, 1990, http://www.nytimes .com/1990/05/19/us/texan-hunkers-down-after-stumbling-on-tongue.html.

When Williams talked about Ann: Steve Daley, "Texans Want 'Other' for Governor," *Chicago Tribune,* November 2, 1990, http://articles.chicagotribune .com/1990-11-02/news/9003310882_1_claytie-williams-republican-clayton -williams-democrat-ann-richards.

"The only way Clayton Williams can win": Michael Holmes, "Ms. Richards Asks: Where's Claytie?," Associated Press, July 5, 1990.

122 *But in July, Williams still led:* David Maraniss, "Letter from Texas; A Vital Need for Intervening Variables," *Washington Post,* July 27, 1990.

"Ann needs to go back": Craig Unger, "How to Beat Bubba," *New Republic,* October 22, 1990.

Williams's ad campaign: Michael Holmes, "Ann Richards, Trailing in Governor's Race, Is Hopeful," Associated Press, August 21, 1990.

"You get the feeling": Roberto Suro, "Texan's Early Strike Makes Democrat Leaders Nervous," *New York Times,* August 4, 1990.

"Richards is struggling": Roberto Suro, "Even for Texas, the Race for Governor Is Rowdy," *New York Times,* September 21, 1990.

"Ann's going to win": Unger, "How to Beat Bubba."

123 *On October 11, at a joint appearance:* Bob Minzesheimer, "Snub Grips Texas Voters; Bitterness Escalates in Gov's Race," *USA Today,* October 15, 1990.

"a crude bumpkin unfit": Roberto Suro, "In Texas, Governor's Race Becomes a Horse Race," *New York Times,* October 29, 1990.

"At no point, Mr. Williams": "Rapist Who Quoted GOP Candidate Gets 50 Years," United Press International, October 29, 1990.

Williams, after all: David Maraniss, "The Texas Two-Step in the Race for Governor," *Washington Post,* October 22, 1990, http://www.washingtonpost .com/archive/lifestyle/1990/10/22/the-texas-two-step-in-the-race-for -governor/a86dac13-487f-48e6-b24c-1a576a182de6/.

All told, Williams's wealth: J. Michael Kennedy Jr., "The Cowboy and the Good Ol' Girl; In an Era of Bland Politics, Claytie Williams and Ann Richards are Brawling, Down and Dirty, in Texas," *Los Angeles Times,* October 21, 1990.

"It is an absolute necessity": Stephen L. Arters, "Millionaire Candidate Says He Paid No Income Tax in 1986," Associated Press, November 2, 1990.

But Williams still said "no": Ross Ramsey, "Bill White and the Politics of Taxes," *Texas Tribune,* March 15, 2010, http://www.texastribune.org/2010/03/15/bill-white-and-the-politics-of-taxes/.

124 *"It's really gonna be neat":* "The 1991 Texas Inauguration of Governor Ann Richards," directed by Cactus and Peggy Davis Pryor, filmed January 15, 1991, Texas Archive of the Moving Image, http://www.texasarchive.org/library/index.php?title=2012_03479, video.

Even better, a Republican poll: David Maraniss, "Texas, It Seems, Has Had Its Fill of Williams; Contrary to Political Tenet, Candidate's Standing in Polls Declines After Massive Media Exposure," *Washington Post,* October 27, 1990.

"It's coming down": Suro, "In Texas, Governor's Race."

125 *"I'll tell you when I didn't pay":* Arters, "Millionaire Candidate Says."

OPEN MOUTH MAY CLOSE: David Maraniss, "Open Mouth May Close a Door in Texas," *Washington Post,* November 2, 1990.

A NEW GAFFE: "A New Gaffe by Candidate in the Texas Governor's Race," *New York Times,* November 3, 1990.

McDonald described it: Mark Langford, "Williams' Admission of Not Paying Income Tax Could Be 'Final Blow,'" United Press International, November 3, 1990.

"This is risky": Thomas Ferraro, "In Texas, Bush Dials for GOP Votes," United Press International, November 4, 1990.

126 *According to the book* Storming the Statehouse: Celia Morris, *Storming the Statehouse: Running for Governor with Ann Richards and Dianne Feinstein* (New York: Scribner, 1992), 174.

127 *That said, for Ann to be elected:* Ellen Goodman, "Shattering a Male Image," *Boston Globe,* November 8, 1990.

128 *One was Joan Finney in Kansas:* Cal Thomas, "Don't Give Up Your Stance on Abortion, GOP," Philly.com, July 30, 1991, http://articles.philly.com/1991-07-30/news/25785045_1_pro-life-plank-abortion-issue-pro-life-position.

The other, Oregon's secretary of state: "Governor Barbara Roberts," Oregon Historical Society, http://history.house.gov/People/Detail/20851.http://www.ohs.org/education/focus/governor-barbara-roberts.cfm.

129 *"This was a first ever":* Pryor and Pryor, "The 1991 Texas Inauguration."

10. Nantucket Sleigh Ride

134　*Pat Schroeder, as chairwoman:* Susan Ferraro, "The Prime of Pat Schroeder," *New York Times,* July 1, 1990, http://www.nytimes.com/1990/07/01/magazine/the-prime-of-pat-schroeder.html.

　Even though laws had been changed: National Institutes of Health, "NIH Celebrates a Decade of Progress in Women's Health Research and the Tenth Anniversary of ORWH," news release, September 8, 2000, http://www.nih.gov/news/pr/sep2000/orwh-08.htm.

135　*"The strenuous demands of Court work":* ABC News, *World News Tonight with Peter Jennings,* video, June 27, 1991.

　"the fact that he is black": Jane Mayer and Jill Abramson, *Strange Justice: The Selling of Clarence Thomas* (Boston: Houghton Mifflin, 1994), 21.

136　*When Bush became president:* Ibid., 151.

　He had never written: Ibid., 21.

　"Anyone who takes him on": Ibid., 20.

137　*Later, when Dixon made his support:* Ibid., 208.

　the last thing she wanted: Ibid., 225.

　It was, as Thomas put it: Jane Flax, *The American Dream in Black and White: The Clarence Thomas Hearings* (Ithaca, N.Y.: Cornell University Press, 1998), 22.

138　*The* New York Times *characterized it:* David Margolick, "Questions to Thomas Fall Short of the Mark," *New York Times,* September 15, 1991.

　National Public Radio's legal affairs correspondent: Nina Totenberg, "Thomas Accused of Sexual Harassment," *Weekend Edition,* National Public Radio, October 6, 1991, http://jwa.org/node/18888.

139　*So, on the House side:* Maureen Dowd, "The Thomas Nomination; 7 Congresswomen March to Senate to Demand Delay in Thomas Vote," *New York Times,* October 9, 1991.

141　*Anita Hill herself finally testified:* "Testimony of Anita F. Hill, Professor of Law, University of Oklahoma, Norman, OK," October 11, 1991, http://mith.umd.edu/WomensStudies/GenderIssues/SexualHarassment/hill-thomas-testimony.

142　*As reported in* Strange Justice: Mayer and Abramson, *Strange Justice,* 235.

143 *On October 15, 1991:* R. W. Apple Jr., "The Thomas Confirmation; Senate Confirms Thomas, 52–48, Ending Week of Bitter Battle; 'Time for Healing,' Judge Says," *New York Times,* October 16, 1991, http://www.nytimes. com/1991/10/16/us/thomas-confirmation-senate-confirms-thomas-52-48 -ending-week-bitter-battle-time.html.

147 *Challenging an incumbent:* Mitchell Locin, "Political Vet Dixon Has Unknown Quality," *Chicago Tribune,* February 23, 1992.

And even if you argued: Kathleen Best, "Dixon's Challengers Growing," *Saint Louis Post-Dispatch,* November 20, 1991.

151 *President Bush had emerged:* Gallup, "Presidential Approval Ratings— Gallup Historical Statistics and Trends," http://www.gallup.com/poll/ 116677/presidential-approval-ratings-gallup-historical-statistics-trends .aspx.

152 *"There isn't a person watching":* "In 1992, Clinton Conceded Marital 'Wrongdoing,'" *Washington Post,* January 26, 1992, http://www.washingtonpost .com/wp-srv/politics/special/clinton/stories/flowers012792.htm.

That year, no fewer than thirteen: Patricia Rice, "Women Voters Outraged by Clarence Thomas Hearings; They May Become Factor in Senate Races," *Saint Louis Post-Dispatch,* March 15, 1992.

Ruth Mandel, a professor: "Ruth B. Mandel," Eagleton Faculty/Staff Bios, Eagleton Institute of Politics, Rutgers University, http://www.eagleton.rutgers .edu/facultystaff/mandel.php.

"That all-white, male Judiciary Committee": Rice, "Women Voters Outraged."

153 *"If the Senate had done its job":* Geraldine Baum, "A Shock to the System; Few People Paid Much Attention to Carol Moseley Braun in the Illinois Democratic Primary, but No One Is Ignoring Her Senate Campaign Now," *Los Angeles Times,* March 27, 1992.

Chicago's Rotary Club: Ibid.

She had virtually no media campaign: Patrick E. Gauen and Robert Goodrich, "Senate Candidates Offer Rescue Plans: Dixon Visits Belleville; Hofeld Airs New Ads," *Saint Louis Post-Dispatch,* February 13, 1992.

Campaign manager Kgosie Matthews: Frank James and Thomas Hardy, "Internal Wars Sour Braun's Campaign," *Chicago Tribune,* February 26, 1992.

154 *One poll showed Carol:* Patrick E. Gauen, "Braun Insists She Has Chance to Upset Dixon," *Saint Louis Post-Dispatch,* March 4, 1992.

If one believed: Patrick E. Gauen, "Dixon Rivals Clash; Senator Joins Debate as Wide Lead Erodes," *Saint Louis Post-Dispatch,* March 9, 1992.

Until this point: Thomas Hardy, "Braun Closing Gap on Dixon; Hofeld Lags, No Candidate Has Clear Path," *Chicago Tribune,* March 8, 1992.

Outspent more than twenty to one: CBS News Transcripts, "In-Depth Look at Primary Results with Dan Rather," *CBS News Special Report,* March 17, 1992.

"People do like to be with somebody": Frank James, "Braun Isn't Invisible Any Longer," *Chicago Tribune,* March 10, 1992.

Civil rights leader Jesse Jackson: Ibid.

155 *"An investment in social justice":* Carole Ashkinaze, "Braun Has Realistic Shot at Senate Seat," *Chicago Sun-Times,* March 12, 1992.

She had raised only roughly $150,000: Patrick E. Gauen, "Polls Give Dixon Edge in Illinois Senate Contest," *Saint Louis Post-Dispatch,* March 8, 1992.

156 *"Senator Alan Dixon is in the fight":* CBS News, "In-Depth Look."

According to the exit polls: Ibid.

Moreover, this phenomenon: Steve Johnson, "Force That Lifted Braun Carries a Wave of Women to Victory," *Chicago Tribune,* March 19, 1992.

157 *According to Judy Erwin:* Ibid.

In the races for Cook County judgeships: Ibid.

11. The Year of the Woman

162 *One of the most interesting races:* Michael DeCourcy Hinds, "The 1992 Campaign Woman in the News: Lynn Hardy Yeakel; Skillful Political Novice," *New York Times,* April 29, 1992, http://www.nytimes.com/1992/04/29/us/the-1992-campaign-woman-in-the-news-lynn-hardy-yeakel-skillful-political -novice.html.

"To have ninety-eight men and two women": Lynn Yeakel, *CNN News,* interview, April 28, 1992.

163 *In March, about a month:* John M. Baer, "Poll Shows Yeakel Is 'Catching Fire,'" Philly.com, April 22, 1992, http://articles.philly.com/1992-04-22/ news/26002611_1_catherine-ormerod-lynn-yeakel-mark-singel.

Then, just three days later: "Yeakel Takes Lead in Democratic Senate Race; Singel 'Running Scared,'" Newswire, April 24, 1992.

Similarly, in Washington State: "Brock Adams Quits Senate Race Amid Sex Misconduct Allegations," *New York Times,* March 2, 1992, http://www.nytimes .com/1992/03/02/us/brock-adams-quits-senate-race-amid-sex-misconduct -allegations.html.

164   *"You're just a mom":* Patty Murray, "No Child Left Behind: Murray Outlines Priorities, Calls for Bipartisan Fix to Broken Law," news release, January 13, 2015, http://www.murray.senate.gov/public/index.cfm/2015/1/video-audio -no-child-left-behind-murray-outlines-priorities-calls-for-bipartisan-fix-to -broken-law.

"Look out world!": Roxanne Roberts, "Flexing Their Electoral Muscle," *Washington Post,* June 5, 1992.

"It's a hundred years later": Jean Godden, "Second Oldest Profession Gets a Link with Oldest," *Seattle Times,* March 25, 1992.

In late April: Mark Matassa, "Women's Fund-Raising Group Adds Murray to List," *Seattle Times,* April 29, 1992.

165   *And in Rep. Mel Levine:* John Balzar, "Democrats' Powerful Alliance Thrives: Waxman-Berman Team; It Spells Clout," *Los Angeles Times,* March 10, 1985, http://articles.latimes.com/1985-03-10/news/mn-25696_1_howard-berman.

In fact, he had assembled: Robert Reinhold, "The 1992 Campaign: California; 2 Women Win Nomination in California Senate Races," *New York Times,* June 3, 1992, http://www.nytimes.com/1992/06/03/us/the-1992-campaign -california-2-women-win-nomination-in-california-senate-races.html.

As the New York Times *put it:* Jane Gross, "The 1992 Campaign: California; Strategies Emerge in Dual Senate Races," *New York Times,* April 11, 1992.

166   *This strategy was effective:* George Skelton, "The Times Poll; Races for Cranston's Seat Tighten Up in Both Parties," *Los Angeles Times,* April 30, 1992.

170   *In Washington:* Barbara Palmer and Dennis Simon, *Breaking the Political Glass Ceiling: Women and Congressional Elections,* 2nd ed. (New York: Routledge, 2008).

Chandler relentlessly attacked: Tim Klass, "GOP Senate Hopeful's Attacks Produce Tie, Gaffes," Associated Press, News Archive, October 24, 1992, http:// www.apnewsarchive.com/1992/GOP-Senate-Hopeful-s-Attacks-Produce-Tie -Gaffes/id-0daf241164f7a779b2bf6c24e0a1b027.

171   *"Dang me, dang me":* Timothy Egan, "In Northwest, Free-for-All Elections," *New York Times,* October 19, 1992, http://www.nytimes.com/1992/10/19/us/ in-northwest-free-for-all-elections.html.

12. Reversal of Fortune

177 *That became abundantly clear:* Bill Clinton, "Remarks at the Signing of the Family Medical Leave Act," White House Rose Garden, February 5, 1993, Miller Center, University of Virginia, http://millercenter.org/president/clinton/speeches/speech-4562.

It had actually passed: Donna R. Lenhoff and Lissa Bell, "Government Support for Working Families and for Communities: Family and Medical Leave as a Case Study," National Partnership for Women and Families, http://www.nationalpartnership.org/research-library/work-family/fmla/fmla-case-study-lenhoff-bell.pdf.

178 *"This is a very poignant moment":* "Family and Medical Leave Signing Ceremony," February 5, 1993, video, C-SPAN, http://www.c-span.org/video/?37753-1/family-medical-leave-signing-ceremony.

"She was terrifying": Clinton, directed by Barak Goodman, PBS, 2012.

179 *One such episode:* Toni Locy, "For White House Travel Office, a Two-Year Trip of Trouble," *Washington Post,* February 27, 1995, http://www.washingtonpost.com/wp-srv/politics/special/whitewater/stories/wwtr950227.htm.

On June 17, 1993: Hillary Rodham Clinton, *Living History* (New York: Simon & Schuster, 2003), Kindle edition.

180 *On July 20, he was found:* David Von Drehle and Howard Schneider, "Foster's Death a Suicide," *Washington Post,* July 1, 1994, http://www.washingtonpost.com/wp-srv/politics/special/whitewater/stories/wwtr940701.htm.

"I made mistakes from ignorance": Sidney Blumenthal, *The Clinton Wars* (New York: Farrar, Straus & Giroux, 2003), 69.

Similarly, Rush Limbaugh: Ibid., 68.

As Politico *magazine reported:* Michael Kruse, "The Long, Hot Summer Hillary Became a Politician," *Politico,* April 14, 2015, http://www.politico.com/magazine/story/2015/04/hillary-clinton-2016-arkansas-116939.html#ixzz3XZfBihwt.

186 *"the best standards we'd ever had":* Ibid.

It was, as one journalist put it: Clinton, *Living History.*

187 *Even though the Clintons lost money:* Encyclopedia.com, s.v. "Whitewater Trials and Impeachment of a President: 1994–99," last modified 2002, http://www.encyclopedia.com/topic/Whitewater.aspx.

The health-insurance lobby: Natasha Singer, "Harry and Louise Return, with a New Message," *New York Times,* July 16, 2009, http://www.nytimes.com/2009/07/17/business/media/17adco.html.

189 *He filmed an ad:* Don Terry, "Illinois Candidate Transforms Herself," *New York Times,* March 21, 1994, http://www.nytimes.com/1994/03/21/us/illinois-candidate-transforms-herself.html.

Over time, she cultivated: "1990 Gubernatorial General Election Results — California," Dave Leip's Atlas of U.S. Presidential Elections, http://uselectionatlas.org/RESULTS/state.php?fips=6&year=1990&f=0&off=5.

At the time, she was known: Suzanna Andrews, "Arianna Calling," *Vanity Fair,* December 2005, http://www.vanityfair.com/culture/features/1994/11/huffington-1994.

190 *He put forth so little substance:* Sidney Blumenthal, "The Candidate," *The New Yorker,* October 10, 1994, http://www.newyorker.com/magazine/1994/10/10/the-candidate-2.

According to Hazel Richardson Blankenship: Maureen Orth, "Arianna's Virtual Candidate," *Vanity Fair,* November 1994, http://www.vanityfair.com/culture/features/1994/11/huffington-199411.

"He's nonexistent": Glenn F. Bunting, "Huffington Draws Fire, but He's Proud of Slim Record," *Los Angeles Times,* May 4, 1994.

It was the highest amount: Stephen K. Medvic, *Political Consultants in U.S. Congressional Elections* (Columbus: Ohio State University Press, 2001), 133.

A January 1994 poll: "Feinstein Holds Big Lead in Poll, 2-to-1 Advantage over 3 Republican Rivals for Senate Seat," *San Francisco Chronicle,* January 22, 1994.

191 *an executive at Raytheon Corporation:* Debra J. Saunders, "The Clueless Philosopher-Candidate," *San Francisco Chronicle,* September 21, 1994.

The bill saved hundreds of jobs: Ibid.

She was now leading: Thomas Farragher, "Huffington Closing In, Poll Says," *San Jose Mercury News,* May 19, 1994.

194 *A man of many contradictions:* Blumenthal, *Clinton Wars,* 125.

In a televised interview: Clinton, *Living History.*

197 *Karen went from a 9-point lead:* "The 1994 Elections: House of Representatives; Who Won Where: Results in the 435 Races for the House," *New York Times,* November 10, 1994, http://www.nytimes.com/1994/11/10/us/1994-elections-house-representatives-who-won-where-results-435-races-for-house.html.

198 *Having been outspent two to one:* Philip J. Trounstine, "Frenetic Windup
     to Campaigns, Wilson Confident," *San Jose Mercury News,* November 8,
     1994.

     According to the California Opinion Index: "A Summary Analysis of Voting in
     the 1994 General Election," California Opinion Index, http://www.field.com/
     fieldpollonline/subscribers/COI-94-95-Jan-Election.pdf.

13. Leaping and Creeping

201 *When the One Hundred Fourth Congress was seated:* James Risen and Elizabeth
     Shogren, "GOP Contract a Reversal of Decades-Old Federal Policies," *Los
     Angeles Times,* December 18, 1994.

205 *"Here was a president":* Clinton, directed by Barak Goodman, PBS, 2012.

     "As long as they insist": Ibid.

     By the end of the summer: Hillary Rodham Clinton, *Living History* (New York:
     Simon & Schuster, 2003), Kindle edition.

206 *"You've been on the plane":* "Gingrich Comment on Shutdown Labeled
     'Bizarre' by White House," CNN.com, November 16, 1995, http://www.cnn
     .com/US/9511/debt_limit/11-16/budget_gingrich/.

     Ever the professor of history: Alexander Nazaryan, "Newt Gingrich, Crybaby:
     The Famous *Daily News* Cover Explained," January 6, 2012, http://www
     .nydailynews.com/blogs/pageviews/newt-gingrich-crybaby-famous-daily
     -news-cover-explained-blog-entry-1.1637386.

     The photo, when juxtaposed: Clinton, *Living History.*

     CRYBABY, *read the headline:* Nazaryan, "Newt Gingrich, Crybaby."

207 *National parks were closed:* Clinton, *Living History.*

214 *By Friday, just three days:* "The Speaker Steps Down; Excerpts from Phone Call
     About Gingrich's Future," *New York Times,* November 8, 1998, http://www
     .nytimes.com/1998/11/08/us/the-speaker-steps-down-excerpts-from-phone
     -call-about-gingrich-s-future.html.

215 *Hillary turned on* Meet the Press: Clinton, *Living History.*

     Finally, in March 1999: Ibid.

218 *Senate Republicans pressured high-tech companies:* Mike Allen, "GOP Pressures
     Tech Firms to Help Michigan Senator; Lobbyists Feared Bill's Fate Hinged on
     Money for Ads," *Washington Post,* May 16, 2000.

By April 2000, Abraham: "Abraham Announces Fund-Raising Advantage over Stabenow," Associated Press, April 14, 2000.

Americans for Job Security: Dee-Ann Durbin, "Pro-Abraham Group Targets Stabenow in Black Newspapers," Associated Press, May 8, 2000.

Right to Life of Michigan: Peter Marks, "Special-Interest Groups Widening Political Attack Ads," *New York Times,* May 14, 2000.

219 *Before the election was over:* Ibid.

Among the first signs: Ryan Thornburg and Ben White, "Going Negative," *Slate,* June 20, 2000, http://www.slate.com/articles/news_and_politics/net _election/2000/06/going_negative.html.

By mid-July, Abraham: "Abraham Looking Stronger for Re-election in Mich.," *National Journal,* July 13, 2000.

By the end of the month: "Can You Name Your U.S. Senators? New MRG Poll Reveals Only 36 Percent of Michigan Voters Can," Newswire, July 26, 2000.

By September 20, his lead: Greg Pierce, "Inside Politics," *Washington Times,* September 20, 2000.

Soon afterward, a poll: Dee-Ann Durbin, "New Poll Shows Stabenow, Abraham in Dead Heat," *Detroit News,* October 26, 2000.

220 *By October 26, a poll:* Ibid.

Then, a November 2 poll: Jack Colwell, "Poll Finds Narrow Lead for Stabenow over Abraham," *South Bend (Ind.) Tribune,* November 4, 2000.

221   *But a week later:* Scott Charton, "Carnahan's Widow Could Wind Up in the Senate in His Place," Associated Press, October 24, 2000.

223 *By Wednesday, with more than 1.6 million votes:* David Ammons, "Gorton, Cantwell Both Lay Claim to Washington Senate Seat Legislature," Associated Press, November 8, 2000.

A week later, Gorton was still leading: Dionne Searcey and David Postman, "Senate Race Tightens in Latest Ballot Count," *Seattle Times,* November 15, 2000.

14. The Next Piece of the Puzzle

227 *"There will always be":* Craig Unger, *The Fall of the House of Bush: The Untold Story of How a Band of True Believers Seized the Executive Branch, Started*

the Iraq War, and Still Imperils America's Future (New York: Scribner, 2007), 250–51, Kindle edition.

230 *She had become the first woman:* "Again, Making Her-Story?," *National Journal,* November 12, 2002.

232 *To figure out exactly:* Andy Kroll, "Can Harold Ickes Make It Rain for Obama?," *Mother Jones,* September–October 2012, http://www.motherjones .com/politics/2012/07/harold-ickes-priorities-usa-action-obama?page=2.

235 *Rather than complain, Gwen helped establish:* Net Industries, *Brief Biographies,* s.v. "Gwen S. Moore Biography," last modified 2005, http://biography.jrank .org/pages/2923/Moore-Gwen-S.html#ixzz3Wef1XSQ5.

Asserting that the police dismissed: William Celis, "Scrutiny of Police Sought in Milwaukee," *New York Times,* July 28, 1991.

236 *Flynn had already lined up:* "In Like Flynn," *National Journal,* June 21, 2004.

239 *In the last reporting period:* "Moore Takes Fund-Raising Lead in Congressional Race," Associated Press, September 3, 2004.

241 *As political analyst Charlie Cook put it:* "America Coming Together (ACT)," Corson.org, http://www.corson.org/archives/soros/soros15_030111.html.

15. Madam Speaker

243 *Indeed, in the first eight months:* "President Bush Job Approval," Real Clear Politics, http://www.realclearpolitics.com/epolls/other/president_bush_job_ approval-904.html.

Even though he barely squeaked by: Richard W. Stevenson, "Confident Bush Outlines Ambitious Plan for 2nd Term," *New York Times,* November 5, 2004, http://www.nytimes.com/2004/11/05/politics/campaign/05bush.html?_r=0.

246 *As early as 2003, Clinton pollster:* Joshua Green, "The Front-Runner's Fall," *The Atlantic,* September 2008, http://www.theatlantic.com/magazine/ archive/2008/09/the-front-runner-s-fall/306944/.

advocating that Hillary: Susan Estrich, *The Case for Hillary Clinton* (New York: ReganBooks, 2005).

In February 2006, ABC News was asking: "Senator Hillary Clinton Signs on 2008; Possible for Presidential Elections," ABC News Now, video, February 13, 2006.

By this point, the answer: Ibid.

248 *As the* Washington Monthly *put it:* Carl M. Cannon, "Why Not Hillary?,"
Washington Monthly, July–August 2005, http://www.washingtonmonthly.com/
features/2005/0507.cannon.html.

250 *Over the years, Murtha secured:* Craig Smith, "Rep. Murtha's Widow Skirts
Politics for Civic Pursuits," *Pittsburgh (Pa.) TribLIVE,* March 22, 2014,
http://triblive.com/state/pennsylvania/5809663-74/murtha-husband
-center#ixzz3jvYwY8kq.

With overwhelming support: Alan Fram, "California Rep. Pelosi Becomes
Highest-Ranking Woman in Congress," Associated Press, October 11, 2001.

"She understands you put in": Karen Breslau, Eleanor Clift, and Daren Briscoe,
with Holly Bailey, Jonathan Darman and Richard Wolffe, "Rolling with Pelosi,"
Newsweek, October 23, 2006.

251 *According to Rep. Anna Eshoo:* Ibid.

A U.N. report on detainees: Craig Unger, *The Fall of the House of Bush: The
Untold Story of How a Band of True Believers Seized the Executive Branch,
Started the Iraq War, and Still Imperils America's Future* (New York: Scribner,
2007), 340, Kindle edition.

Polls by Newsweek, *Pew Research:* "President Bush Job Approval," Real Clear
Politics.

252 *"I've acquired a lot of new friends":* Adam Nagourney and Robin Toner, "With
Guarded Cheer, Democrats Dare to Believe This Is Their Time," *New York
Times,* October 22, 2006.

"We know there's a hurricane": Susan Page and Richard Wolf, "Republicans
May Be in Path of Political 'Hurricane,'" *USA Today,* April 5, 2006.

Sen. Maria Cantwell: "America Votes, 2006: U.S. Senate/New York,"
CNN.com, http://www.cnn.com/ELECTION/2006/pages/results/states/
NY/S/01/.

16. Down to the Wire

258 *A bit more than two weeks later:* "Hillary Clinton Launches White House
Lid: 'I'm In,'" CNN.com, January 22, 2007, http://www.cnn.com/2007/
POLITICS/01/20/clinton.announcement/index.html?eref=yahoo.

"I am one of the millions of women": Beacon Journal Wire Services, "Clinton
Declares She's In, Vows That 'I'm In to Win'; Democrat Considered to Have
Best Shot to Be First Woman President," *Akron (Ohio) Beacon Journal,* January
21, 2007.

259 *By the end of the first quarter:* "Campaign Finance: First Quarter 2007 FEC Filings," *Washington Post,* 2007.

260 *There were Hillary Clinton nutcrackers:* Hillary Clinton nutcracker, Kaboodle. com, http://www.kaboodle.com/reviews/hillary-clinton-nutcracker-stupid .com.

Babies' bibs sported the slogan: "I Wish Hillary Had Married OJ" infant T-shirt, CaféPress.co.uk, http://www.cafepress.co.uk/+i_wish_hillary_had _married_oj_infant_tshirt,494240597.

There was a Facebook page: Amanda Fortini, "The Feminist Reawakening," *New York Magazine,* April 13, 2008, http://nymag.com/news/features/46011/.

Chris Matthews referred to Hillary: Dodai, "Chris Matthews Has a Sexist History with Hillary Clinton," Jezebel.com, January 15, 2008, http://jezebel .com/345237/chris-matthews-has-a-sexist-history-with-hillary-clinton. Nurse Ratched is the cold, tyrannical nurse in Ken Kesey's *One Flew Over the Cuckoo's Nest,* and Madame Defarge is the villainess in Charles Dickens's *A Tale of Two Cities.*

Other pundits saw fit: Susan Campbell, "Revisiting Carl Bernstein's 'On Clinton's "Thick Ankles," '" *Hartford Courant,* June 18, 2008, http://articles. courant.com/2008-06-18/features/susan0618_1_hillary-clinton-clinton-years -clinton-s-thick-ankles.

"The historians can put aside": Bob Herbert, "The Obama Phenomenon," *New York Times,* January 5, 2008, http://www.nytimes.com/2008/01/05/ opinion/05herbert.html?pagewanted=print.

Liberal blogger Ezra Klein: Ezra Klein, "Obama's Gift," *American Prospect,* January 4, 2008, http://prospect.org/article/obamas-gift.

Talk-show icon Oprah Winfrey: Ben Smith and David Paul Kuhn, "Messianic Rhetoric Infuses Obama Rallies," *Politico,* December 8, 2007, http://www .politico.com/news/stories/1207/7281.html.

261 *As Joshua Green pointed out:* Joshua Green, "The Front-Runner's Fall," *The Atlantic,* September 2008, http://www.theatlantic.com/magazine/ archive/2008/09/the-front-runner-s-fall/306944/.

262 *Instead, delegates are selected:* "Election 2008: Iowa Caucus Results," *New York Times,* http://politics.nytimes.com/election-guide/2008/results/states/ IA.html.

When the lobbying ended: "Iowa Caucuses 101: Arcane Rules Have Huge Impact on Outcome," CNN.com, January 3, 2008, http://www.cnn.com/2008/ POLITICS/01/03/iowa.caucuses.101/.

263 *When the votes were counted:* "Election 2008: Iowa Caucus Results," *New York Times,* accessed October 16, 2015, http://politics.nytimes.com/election -guide/2008/results/states/IA.html.

"If Hillary doesn't stop Obama": Patrick Healy and Marjorie Connelly, "Bad Night for Status Quo on Right and Left; Victories by Outsiders Leave No Clear Path to Party Nominations," *New York Times,* January 4, 2008.

But the New Hampshire primary: "New Hampshire Democratic Primary," Real Clear Politics, January 8, 2008, http://www.realclearpolitics.com/epolls/2008/ president/nh/new_hampshire_democratic_primary-194.html.

The next day, January 5: "The Democratic Debate in New Hampshire," *New York Times,* January 5, 2008, http://www.nytimes.com/2008/01/05/us/ politics/05text-ddebate.html?pagewanted=all.

Four candidates made the cutoff: Patrick Healy and Jeff Zeleny, "At Debate, Two Rivals Go After Defiant Clinton," *New York Times,* January 6, 2008.

264 *"Making change is not about":* "Democratic Debate in New Hampshire."

"Well, that hurts my feelings": Ibid.

265 *"I couldn't do it":* "Hillary Clinton Tears Up During Campaign Stop," uploaded January 7, 2008, YouTube video, https://www.youtube.com/ watch?v=6qgWH89qWks.

"Can Hillary cry her way": Maureen Dowd, "Can Hillary Cry Her Way Back to the White House?," *New York Times,* January 9, 2008.

"That crying really seemed genuine": Ibid.

266 *When I called my friends:* Pat Cunningham, "Obama Tsunami Hitting New Hampshire?," rrstar.com, January 8, 2008, http://blogs.e-rockford.com/ applesauce/2008/01/08/obama-tsunami-hitting-new-hampshire/ #axzz3eTzRsb00.

and saying he would win: "The Hotline Blogometer; Independents Flex Their Muscles," *National Journal,* January 8, 2008.

267 *She had carried the women's vote:* "Clinton Wins Back Women, Narrowly Takes New Hampshire," CNN.com, January 9, 2008, http://www.cnn. com/2008/POLITICS/01/08/democrat.results/index.html?eref=yahoo.

In March, Hillary won Ohio and Rhode Island: "Election 2008: Ohio Primary Results," *New York Times,* http://politics.nytimes.com/election-guide/2008/ results/states/OH.html; "Election 2008: Rhode Island Primary Results," *New York Times,* http://politics.nytimes.com/election-guide/2008/results/states/ RI.html.

17. EMILY's List 2.0

271 *"I can still be disappointed":* Emily Pierce, "History, Not Her Story, in Denver," *Roll Call,* August 26, 2008, http://www.rollcall.com/issues/54_21/-27574-1 .html.

272 *No event was more difficult:* Michael Falcone, "Mrs. Obama Makes a Pitch for Women Voters," *New York Times,* June 20, 2008, http://thecaucus.blogs. nytimes.com/2008/06/20/mrs-obama-makes-a-pitch-for-women-voters/ ?_r=0.

    At the Democratic National Convention: Mary Katharine Ham, "Emily's List Gala to Propel Hillary into Tuesday Speaking Slot," *Washington Examiner,* August 25, 2008, http://www.washingtonexaminer.com/emilys-list-gala-to -propel-hillary-into-tuesday-speaking-slot/article/10391.

274 *As an author, attorney, and daughter:* Glenn Blain and Corky Siemaszko, "Caroline Kennedy Meets Syracuse Politicians as She Campaigns for Hillary Clinton's Senate Seat," *New York Daily News,* December 18, 2008, http://www .nydailynews.com/news/politics/caroline-kennedy-meets-syracuse- politicians-campaigns-hillary-clinton-senate-seat-article-1.355375.

    New York mayor Mike Bloomberg: Corky Siemaszko, "Why Caroline Kennedy's Senate Bid Flamed Out," *U.S. News and World Report,* January 23, 2009, http://www.usnews.com/news/articles/2009/01/23/why-caroline-kennedys -senate-bid-flamed-out.

    But beyond her family background: Erin Einhorn and David Saltonstall, "Records Show Caroline Kennedy Failed to Cast Her Vote Many Times Since 1988," *New York Daily News,* December 19, 2008, http://www.nydailynews .com/news/politics/records-show-caroline-kennedy-failed-cast-vote-times -1988-article-1.355381.

    On January 22, Kennedy announced: Keith B. Richburg, "Kennedy Drops Bid for Seat in Senate," *Washington Post,* January 22, 2009, http://www .washingtonpost.com/wp-dyn/content/article/2009/01/21/AR2009012103205 .html.

    After another conversation: "Clinton's Senate Successor Is Rep. Kirsten Gillibrand: NY Station," *Top of the Ticket* (blog), latimes.com, January 22, 2009, http://latimesblogs.latimes.com/washington/2009/01/ny-tv-station-s .html.

276 *Tester won by the thinnest:* Ben Terris, "Tester, Rehberg, Too Close to Call," *National Journal,* November 7, 2012, http://www.nationaljournal.com/ congress/tester-rehberg-too-close-to-call-20121107.

Al finally emerged victorious: Manu Raju and Josh Kraushaar, "Norm Coleman Concedes Minnesota Senate Race to Al Franken," *Politico,* July 1, 2009, http://www.politico.com/news/stories/0609/24383.html.

279 *Having been named America's Sexiest Man:* Ashley Womble, "Senator Scott Brown Posed Nude for Cosmo," *Cosmopolitan,* September 20, 2012, http://www.cosmopolitan.com/politics/news/a10586/scott-brown-nude-in-cosmo/.

Even though the national Tea Party: Levi Russell, "Tea Party Express Spending in Support of Scott Brown Surpasses $300,000+!," *Free Republic,* January 18, 2010, http://www.freerepublic.com/focus/news/2431678/posts/.

In the end, Brown won: "GOP's Scott Brown Wins Mass. Senate Race," CBSNews.com, January 19, 2010, http://www.cbsnews.com/news/gops-scott-brown-wins-mass-senate-race/.

280 *On the bright side:* Associated Press, "Terri Sewell Becomes First Black Woman Elected to U.S. House from Alabama," AL.com, November 2, 2010, http://blog.al.com/wire/2010/11/ap_terri_sewell_becomes_first.html.

Democrats lost a staggering sixty-three seats: Jesse Byrnes, "DCCC Head: 'Probably' Won't Win House," *Ballot Box* (blog), TheHill.com, November 3, 2014, http://thehill.com/blogs/ballot-box/house-races/222629-dccc-head-environment-for-dems-worse-than-2010.

281 *Not long afterward, in September 2010:* "President Obama Names Elizabeth Warren Assistant to the President and Special Advisor to the Secretary of the Treasury on the Consumer Financial Protection Bureau," White House press release, September 17, 2010, https://www.whitehouse.gov/the-press-office/2010/09/17/president-obama-names-elizabeth-warren-assistant-president-and-special-a.

283 *In response, Rush Limbaugh:* Maggie Fazeli Fard, "Sandra Fluke, Georgetown Student Called a 'Slut' by Rush Limbaugh, Speaks Out," *Washington Post,* March 2, 2012, http://www.washingtonpost.com/blogs/the-buzz/post/rush-limbaugh-calls-georgetown-student-sandra-fluke-a-slut-for-advocating-contraception/2012/03/02/gIQAvjfSmR_blog.html.

Needless to say, it was abortion: U.S. House of Representatives, Committee on Energy and Commerce, "Ranking Members Waxman and DeGette Urge Chairman Stearns to Reconsider Planned Parenthood Investigation," news release, September 27, 2011, http://democrats.energycommerce.house.gov/?q=news/ranking-members-waxman-and-degette-urge-chairman-stearns-to-reconsider-planned-parenthood-inves.

284 *Former senator Rick Santorum:* Igor Volsky, "Rick Santorum Pledges to Defund

Contraception: 'It's Not Okay, It's a License to Do Things,'" ThinkProgress, October 19, 2011, http://thinkprogress.org/health/2011/10/19/348007/rick -santorum-pledges-to-defund-contraception-its-not-okay-its-a-license-to-do-things.

Mitt Romney, the eventual Republican: Robin Abcarian, "GOP Debate: Mitt Romney Grows Foggy on Contraception," *Los Angeles Times,* January 07, 2012, http://articles.latimes.com/2012/jan/07/news/la-pn-mitt-romney-grows-foggy -on-contraception-question-20120107.

Jay Townsend, a spokesman: Melissa Jeltsen, "Jay Townsend, GOP Spokesman: 'Let's Hurl Some Acid at Those Female Democratic Senators,'" *Huffington Post,* May 31, 2012, http://www.huffingtonpost.com/2012/05/31/jay-townsend -nan-hayworth-acid-war-on-women_n_1560693.html.

Janice Daniels, the Republican mayor: Kate Abbey-Lambertz, "Janice Daniels, Michigan Mayor Who Was Recalled After Anti-Gay Slurs, Is Back," *Huffington Post,* August 14, 2013, http://www.huffingtonpost.com/2013/08/14/janice -daniels-troy-michigan-city-council_n_3757459.html.

285 *We could point out that in the fifty states:* "States Enact Record Number of Abortion Restrictions in 2011," Guttmacher Institute, January 5, 2012, http://www.guttmacher.org/media/inthenews/2012/01/05/endofyear .html.

286 *Then, nearly eight years later:* Rod Ohira, "Lieutenant Governor Reflects on the 'Bookends' of Her Life," *Honolulu Star-Bulletin,* May 8, 1999, http://archives .starbulletin.com/1999/05/08/news/story5.html.

Money was so tight: "Meet Mazie Hirono," Mazie for Hawaii: U.S. Senate, http://www.mazieforhawaii.com/who/biography.

287 *In 1980, she was elected:* Bob Dye, "Hirono Faces Uphill Battle in Upcoming Governor's Race," *Honolulu Advertiser,* March 4, 2001, http://the .honoluluadvertiser.com/2001/Mar/04/34opinion16.html.

18. Women Make It Work

289 *Tammy Baldwin joined the fight:* Tammy Baldwin, "On Equal Pay Day U.S. Senator Tammy Baldwin Calls for Support of Paycheck Fairness Act," press release, April 9, 2013, http://www.baldwin.senate.gov/press-releases/on -equal-pay-day-us-senator-tammy-baldwin-calls-for-support-of-paycheck -fairness-act.

In the House, Kyrsten Sinema (D-AZ): Kyrsten Sinema, "ICYMI: Valley Tech

Manufacturer: Sinema's Jobs Bill Is Tremendously Helpful," press release, April 11, 2013, http://sinema.house.gov/index.cfm/press-releases?ID= 584babaf-b32d-4296-a290-e7ba6ade532c.

Cheri Bustos (D-IL) introduced: Cheri Bustos, "Congresswoman Cheri Bustos Introduces Legislation to Prevent Student Loan Interest Rates from Doubling in July," press release, April 9, 2013, http://bustos.house.gov/press-and-media/ press-releases/congresswoman-cheri-bustos-introduces-legislation-to -prevent-student.

Likewise, Louise Slaughter (D-NY) and Gwen Moore (D-WI): Gwen Moore, "Reps. Moore, Conyers, and Slaughter Introduce VAWA That Protects LGBT, Native American, Immigrant, Campus, and Sex Trafficking Victims: Bipartisan Senate Version Of VAWA Deserves a Vote in House," press release, February 26, 2013, http://gwenmoore.house.gov/press-releases/reps-moore-conyers -and-slaughter-introduce-vawa-that-protects-lgbt-native-american -immigrant-campus-and-sex-trafficking-victims-bipartisan-senate-version -of-vawa-deserves-a-vote-in-house/.

In its efforts to defund: Kirsten Appleton and Veronica Stracqualursi, "Here's What Happened the Last Time the Government Shut Down," ABC News, November 18, 2014, http://abcnews.go.com/Politics/heres-happened-time -government-shut/story?id=26997023.

290 *For more than two weeks:* Brad Plumer, "Absolutely Everything You Need to Know About How the Government Shutdown Will Work," *Washington Post,* September 30, 2013, http://www.washingtonpost.com/blogs/wonkblog/ wp/2013/09/30/absolutely-everything-you-need-to-know-about-how-the -government-shutdown-will-work/.

The old boys' network: Jay Newton-Small, "Women Are the Only Adults Left in Washington," *Time,* October 16, 2013, http://swampland.time.com/2013/10/16/ women-are-the-only-adults-left-in-washington/.

As Time *titled an article:* Ibid.

"I think what our group did": Jonathan Weisman and Jennifer Steinhauer, "Senate Women Lead in Effort to Find Accord," *New York Times,* October 14, 2013, http://www.nytimes.com/2013/10/15/us/senate-women-lead-in-effort -to-find-accord.html?_r=0.

291 *Soon, the GOP trio began working:* Ibid.

"Leadership, I must fully admit": Laura Bassett, "Men Got Us into the Shutdown, Women Got Us Out," *Huffington Post,* October 20, 2013, http:// www.huffingtonpost.com/2013/10/16/shutdown-women_n_4110268.html.

This meant that Patty Murray: Paul Ryan and Patty Murray, "Murray and Ryan Introduce Bipartisan Budget-Conference Agreement," press release, December 10, 2013, http://budget.house.gov/news/documentsingle.aspx?DocumentID= 364030.

Instead of trying for "a grand bargain": Alex Rogers, "Budget Bill Heads Toward Senate Approval," *Time,* December 17, 2013, http://swampland.time.com/2013/ 12/17/budget-bill-heads-toward-senate-approval/.

"If we didn't get a deal": Patty Murray, "Murray-Ryan Budget Agreement Passes Senate," press release, December 18, 2013, http://www.murray.senate .gov/public/index.cfm/2013/12/murray-ryan-budget-agreement-passes-senate.

"We did it because we listened": Tom Raum, "Omnibus Bill Passage Brings Rare Truce in Fiscal Wars," *Huffington Post,* January 17, 2014, http://www .huffingtonpost.com/2014/01/17/omnibus-bill-passage_n_4616251.html.

292 *In short, it was far more amenable:* Jennifer Steinhauer, "Farm Bill Reflects Shifting American Menu and a Senator's Persistent Tilling," *New York Times,* March 8, 2014, http://www.nytimes.com/2014/03/09/us/politics/farm-bill -reflects-shifting-american-menu-and-a-senators-persistent-tilling.html?_r=0.

293 *As a result, in February 2014:* Ibid.

294 *"The Committee has reviewed":* Eugene Robinson, "Lessons of Benghazi," *Washington Post,* January 17, 2014.

Nevertheless, Sen. Rand Paul (R-KY): Dana Milbank, "Ted Cruz's Never-Ending Ego Trip," *Washington Post,* February 16, 2014.

Like the Senate, on July 31, 2014: Dana Milbank, "GOP Finally Gets Real About Fake News," *Washington Post,* August 14, 2014.

By May 2014, after there had already been: John Bresnahan, Lauren French, and Jake Sherman, "Mission Improbable: Trey Gowdy Gets into Benghazi," *Politico,* May 4, 2014, http://www.politico.com/story/2014/05/mission-improbable -trey-gowdy-gets-into-benghazi-106326.html.

no fewer than 206 Republicans: Dylan Scott, "206 House GOPers Fighting for 7 Seats on the Benghazi Committee," *Talking Points Memo,* May 7, 2014, http://talkingpointsmemo.com/livewire/206-house-republicans-benghazi -committee.

In 2014, they lost thirteen seats: John Woolley and Gerhard Peters, "Seats in Congress Gained/Lost by the President's Party in Mid-term Elections," American Presidency Project, http://www.presidency.ucsb.edu/data/mid -term_elections.php.

296 *At this point, Republicans in Congress:* David Brock, "Brock: What's Behind the House Select Committee on Media Matters?," *Media Matters for America* (blog), June 22, 2015, http://mediamatters.org/blog/2015/06/22/brock-whats -behind-the-house-select-committee-o/204088.

In more than nine hours: Ibid.

but more than 160 questions: Elijah E. Cummings, "Blumenthal Deposition Transcript Ready for Release," press release, June 19, 2015, http://democrats. benghazi.house.gov/news/press-releases/blumenthal-deposition-transcript- ready-for-release.

As Adam Smith (D-WA), a member: Tim Mok, "Blumenthal: Free My Benghazi Testimony," *Daily Beast,* June 19, 2015, http://www.thedailybeast. com/articles/2015/06/19/blumenthal-free-my-benghazi-testimony.html.

298 *see exactly how far:* Derek Willis, "G.O.P. Women in Congress: Why So Few?," *New York Times,* June 1, 2015, http://www.nytimes.com/2015/06/02/upshot/ gop-women-in-congress-why-so-few.html.

299 *Today, 14 out of 44:* "History of Women in the U.S. Congress," Center for American Women and Politics, Rutgers University, http://www.cawp.rutgers. edu/history-women-us-congress.

300 *Even though there were twenty women:* "Women in Congress: Leadership Roles and Committee Chairs," Center for American Women and Politics, Rutgers University, http://www.cawp.rutgers.edu/sites/default/files/resources/conglead. pdf.

But, thanks largely to Senator Mikulski's efforts: Barbara Mikulski, "Mikulski Spotlights Need to Provide Robust and Reliable Federal Investment in Medical Innovation," press release, April 3, 2014, http://www.mikulski.senate.gov/ newsroom/press-releases/mikulski-spotlights-need-to-provide-robust-and- reliable-federal-investment-in-medical-innovation.

INDEX